景观与城市转变
landscape transformations

urbicus
宇比库斯事务所 设计作品专辑

Green Vision 绿色观点·景观设计师作品系列

本系列图书为法国亦西文化公司(ICI Consultants/ICI Interface)的原创作品，原版为法英文双语版。
This series of works is created by ICI Consultants/ICI Interface, in an original French-English bilingual version.

法国亦西文化 ICI Consultants 策划编辑

总企划 Direction：简嘉玲 Chia-Ling CHIEN
协调编辑 Editorial Coordination：尼古拉·布里左 Nicolas BRIZAULT
英文翻译 English Translation：艾莉森·库里佛尔 Alison CULLIFORD
中文翻译 Chinese Translation：陈庶 Shu CHEN
中文校阅 Chinese Proofreading：简嘉玲 Chia-Ling CHIEN
版式设计 Graphic Design：维建·诺黑 Wijane NOREE
排版 Layout：卡琳·德拉梅宗 Karine de La MAISON

绿色观点·景观设计师作品系列
green vision

景观与城市转变
landscape transformations

urbicus

jean-marc gaulier

宇比库斯事务所 设计作品专辑

让马克·高里耶

辽宁科学技术出版社

芒特拉若利的露天绿地剧场，2006年。

Open-air theatre in Mantes-la-Jolie, 2006.

foreword 前言

景观犹如人类与自然之间的媒介，是人类对其周围环境所作所为的综合呈现：景观是人类改造自然的持续转变过程所产生的结果。

土地的过度开发、人类对自然的掠夺、经济优先于社会的取向，这一切使得景观环境遭到破坏。景观展示了社会政策的质量，在许多失败的社会措施中，"景观艺术"经常是缺席的。

为了深层次地改造社会政策，我们必须先改变面对景观的态度。

可持续性发展就是试图达成社会、环境和经济平衡的社会政策。景观是这些平衡政策在土地上所呈现的面貌，而可持续性也正是景观的特点。

具有全面性与可持续性的政策是一种循序转变的方法。

无论是在大自然、乡村或是城市环境中，景观都是预先存在的。景观设计师并没有创造景观，而是将其展现出来、提升其价值、使其持久存在或者进行转化。

景观不属于我们，它是土地的表情、是"国家公有的遗产"（法国城市规划法第110条款）。景观方案并不是我们设计者的游戏场地，而是结合了各种必要元素的结果：一个被认同的方案、一块展现出特色的基地、一个社会需求的体现，以及一个在掌控之中的转变。

As a medium between Man and nature, the landscape is a synthesis of human action on the environment. The landscape is the result of the perpetual process of Man's transformation of nature.

Sorry landscapes are born from the excessive pressure exerted on our land, from Man's predation on nature, from giving the economy prevalence over society. The landscape is a gauge of the quality of our projects for society. And all too often they fall short, with the "landscape arts" too often serving as an alibi.

Transforming our attitude to landscape is therefore necessary and indispensable for transforming what we want for society.

Sustainable development is the societal project that seeks to balance the social, the environmental and the economic. Landscape architecture is the result of these same equilibria applied to land areas. Sustainability is the very essence of landscape architecture.

The global and sustainable project is a means of transformation.

Whether it is natural, rural or urban, the landscape pre-exists. The landscape architect doesn't create landscapes, he reveals them, gives them back their value, sustains them or transforms them.

The landscape does not belong to us, it is the way the Land Area expresses itself, and this Land Area (or territoire in French) is a "common heritage of the Nation" (article L.110 of the French town planning code). Landscape projects are not our playing fields, but are the result of a shared project, of sites revealed, of fulfilling a social commission, and of responsible transformation.

封面照片：安赞扎克-洛克瑞斯特的布拉维河畔（建筑设计：DDL Architectes）。

Photo of the cover: The Blavet riverside fringes at Inzinzac-Lochrist (buildings: DDL Architectes).

当我们着手一个方案的时候，我们不追求带有自恋色彩的创新、专业上的独树一帜以及能够引发新闻价值的新事物，而是试图建立一个与项目紧密结合的设计步骤，以期通过一个专注而机智的方法来分析场地、清楚地辨识出重要课题，并为此提出答案。

这些答案的持久性意味着执行方法上的精简、面对基地和其居民的谦逊态度、对自然现象的考量、对日常性策略的实施，和对个人风格表现的规避。

在方案设计阶段，景观就是一块展现出特色的基地，通过各种步骤而逐步完善成形：精心地规划功能、建立层级分明的条件、拟定明智的行动措施、建立融洽的合作关系，从而得出一个获得认同的工作结果。

一个方案的质量取决于：最初的探询和提问是否适切而明智、结论是否严谨而合理，以及转变过程是否尊重基地特性并且具有持久性。

景观规划是一项可持续发展的城市规划，在其中，景观是有生命的，是舒适、有品质的生活环境，是受到重视和改善的城市环境和自然环境，也是一种节约且具有平衡性的土地整治方式。

我们希望看到土地规划方法的转变，在这些方法中，景观不是由农耕者、生态学家、建筑师、城市规划师、工程师、企业主以及他们各自所属的政府行政单位独自决定而不加商议的结果，而是来自一种整体性的管理，积极通过可持续发展的景观来建立具有全面性的土地规划。

We avoid approaching the project from the angle of narcissistic creativity, professional originality or media-friendly novelty. We insist on a project methodology that aims to produce coherent solutions to clearly identified problems through an attentive and inventive approach to the sites.

The sustainability of these answers depends on a certain economy of means, a modesty vis-à-vis the site and its inhabitants, taking natural phenomena into account, putting in place ordinary strategies and rejecting an excess of personal expression.

A landscape project means a site revealed and informed by drawing up a programme, putting challenges in order of priority, distinct action scenarios, listening to partners and working together.

The quality of a project therefore comes from asking pertinent questions, finding coherent solutions, and the sustainability of a transformation process.

Landscape architecture means a sustainable form of town planning where the landscape is liveable, a viable quality of living, an urban and natural landscape that makes the most of its assets, a land area development that is economical and balanced.

We dream of seeing land area development methods transformed, so that the landscape would not result from the individual and non-consultative impulse of the farmer, the ecologist, the architect or town planner, the engineer and the contractor and their respective ministries, but be protected by an administrative body whose vocation would be to produce an overall development of the land area through a sustainable landscape.

我们为了创造这样一种景观而努力工作着，它是一个共享的整体规划，是一个平衡社会的体现，它从此不再是与周围环境毫无呼应的各种景观的并列结果，例如那些被冠以历史或自然遗产头衔的保护景观、为经济活动而牺牲的景点，或是被遗忘、被忽略、被认为"无趣"的风景。

新景观的创造必定经过对现有景观的仔细分析。对我们在土地上进行的有关经济、环境和社会的活动采取严谨的现状评估，是创造21世纪可持续性发展景观的本质。

这本图面丰富的方案汇编并不是专题讨论，也不是传记，它仅仅是为了展现我们在景观规划与设计中所运用的多种方法，通过回顾一些经过挑选的方案来见证：景观能够以一种"可持续"的方式回应土地整治过程中所出现的种种课题。

这本书呈现了一个团队的工作，希望借此展示出景观设计师在专业实践上的多样性。这并不是为了团队本身，而是为了改善人们对景观学科的整体认识。这个景观学科可以作为我们的社会政策和方案的整合者，然而，在21世纪的法国，这方面的意识并不足够。一个建筑师-景观设计师的专业实践，也是景观师作为"园丁"和建筑师作为"建筑工人"的实践。

We strive for the landscape no longer to be a no-comment juxtaposition of landscapes protected as historical or natural heritage, landscapes sacrificed for the benefit of economic activity and forgotten landscapes, overgrown or "useless", but instead a coherent shared project where the landscape reflects a balanced society.

The invention of new landscapes has to involve the inventory of the old ones. Only through a critical study of the existing landscapes, considering the economic, environmental and social practices of the past, can we produce sustainable landscapes for the 21st century.

With this plurigraphic collection – neither a monograph nor a biography – we simply wish to testify to the multitude of methods of acting, and to review chosen projects to illustrate that the landscape is a pertinent way of considering the problems of land area development.

This book shows the work of a team that wishes to demonstrate the diversity of landscape architects' practices, not for themself, but to show the value of landscape architecture. This disciplinary field is not sufficiently recognised as such in 21st-century France, but could be a unifying force for our societal projects. The practice of architect-landscape architect, or of landscape "gardener", should be recognised etymologically, as "civic builder" is in architecture.

芒特拉若利的露天绿地剧场，一旁为修道院附属教堂，2006年。 Open-air theatre in Mantes-la-Jolie, view of the abbey, 2006.

contents 目 录

大地建筑
architecture of the land — 010

空间的分享与共享的城市
sharing space and the shared city — 012

整体的方案是共创的方案
a project as a whole, a project together — 056

"自然城市性"
the "naturbanity" — 076

"正能源城市"
a "positive city" — 078

新景观经济
a new economy of landscape — 110

方案索引
projects index — 130

事务所简介
biography-agency — 132

致谢
contributions — 133

版权说明
credits — 134

大地建筑

architecture of the land

景观是一个有生命的建筑,是人类历史和自然历史、环境层理和社会实践的重叠,其方案构思则是一场永恒的运动。它是一个空心的体量,是我们的社会在自然当中所展现的模样,是深奥环境里面的具象组织,是将纸上地籍图转化为具体地块的临界建筑,是由我们种种活动所产生的无止境的马赛克拼组,是众多未经协调的方案的并置。

作为我们看见的实物的总和,景观具有美学与感性的面向,然而,我们希望像社会政策一样使其客观化,并且以一种组构功能、生态和经济的方式来处理它。景观是一种"大地建筑"。

景观是由我们自身的特性所组成的。它犹如我们社会的样貌,是我们的社会表征。它也犹如我们私人地址所在的场所,界定了他人对我们的目光,以及我们对外展示的形象。

The landscape is a living architecture, a superimposition of human history and natural history, of environmental stratifications and social practices whose project is in perpetual motion. It is a volume in a cavity, the moulding of nature by our society, a tangible organisation of the impalpable environment, the architecture of boundaries that transforms the invisible cadastre into visible plots, the infinite mosaic resulting from our practices, and the juxtaposition of several uncoordinated projects.

The landscape, as a collection of things seen, has an aesthetic and tangible dimension, but we want to objectify it as a social project and to approach it as a way of architecting uses, ecology and economy. The landscape is an "architecture of the land".

The landscape composes our identities. As representative of our societies, it is our social identity. As our personal address, it defines others' view of ourselves and the image we send back to them.

毁坏景观，就好像毁损我们的容貌、降低我们的身份、轻视历史、切断我们的根基，并且使自然环境失去平衡。遭到破坏的景观是一个无法生存的环境。工业化的农业是一种不公正且带有卫生事故风险的乡村经济模式。生活在一个缺乏合宜景观的空间里，犹如居住在一个没有地址的街区，也等同于是社会的死亡。

景观能够整合多种事物，例如：空间功能组织、我们与他人产生关系的场所、我们的交通空间、我们的食物品质以及我们在生态环境中进行经济活动所带来的后果。

作为景观设计师，我们的工作便是构思这个大地建筑，这个唯一能将功能、时间与自然现象同时纳入考量的建筑。

Mistreating the landscape is a way of disfiguring ourselves, of devaluing our identity, scorning history, mutilating our roots and unbalancing natural environments. A debased landscape cannot be lived in. Industrialised agriculture is a one-sided rural economy and an accident waiting to happen. An anonymous housing estate is a ghetto where an absence of landscape to enhance the living environment is synonymous with social death.

Landscape structures the uses of a land area, the space in which we conduct our relationships with others, the reach of our means of transport, the quality of our food and the consequences of our economic practices on the environment.

As landscape architects, we are the builders of this architecture, which is the only kind that integrates the understanding of uses, time and natural phenomena.

瓦勒市多媒体图书馆的喷泉花园。
（建筑设计：David Cras Architecte）

Fountain garden for the Vallet mediatheque
(Building: David Cras Architecte).

Sharing space and the shared city
空间的分享与共享的城市

公共空间是一个能够组织城市和赋予城市品质的设施。城市消费主义简化地将城市功能并列和堆叠，并且污染土地、使土地贫瘠化。20世纪的城市受到"汽车魔力"的影响，是以汽车用地大小来确定城市尺度的。

我们现在必须以不同的方式面对城市中的汽车问题，使汽车从城市的空间和人们的思想中消失。但我们的经济发展不允许机动车的完全消失，因此汽车共享、停车场互惠、短途出行时以柔性交通(步行、滑轮、自行车等)取代汽车、城市公共交通的改善、无污染汽车的使用，这些措施都将为城市景观带来改革性的变化。

电动汽车将使得今日环境质量低落的高速公路周边地区无更为城市化。如果人们不再强调速度，公路可以改变为街道，街道则成为汽车和步行者共享的交往空间。如果汽车的拥有不再是社会身份的象征，由汽车共享以及停车场的减少而产生的空间，便可以服务于其他城市功能。

室外空间以它的位置、尺度和界定方式，将城市中的建筑展现出来。这个室外空间的形式语汇应该与建筑的形式语汇形成互补。边界和围墙塑造出公共空间和私人空间的过渡，植物则以不同的高度层次在行人和建筑物之间形成了一个中间尺度。

犹如大地建筑的景观，表达出了人们对空间均衡分享的新需求与新界线。

Public space is an urban facility that organises and defines the city. The consumerist city, juxtaposing and piling up functionalities, pollutes and impoverishes the land area. In the grip of its fascination with the motor car, the 20th-century based the dimensions of its cities on the width of vehicles.

We must now take a new view of the place of the car in cities by "de-vehicling" both the space and our preconceptions. From an economic point of view, we cannot envisage getting rid of cars completely, but the shared car, pooled parking spaces, the replacement of the car on short journeys by soft transport, the improvement of public transport and clean cars are among the hypotheses that will cause a revolution in the urban landscape.

The electric car will make it possible for the borders of motorways, which cannot be lived on today, to become urban. If speed is not an issue, the road can become a street, a place where cars and pedestrians cohabit. If the possession of a car is not a sign of social status, car pooling and a reduction in the number of parking spaces will generate room for other uses.

Through its siuation, its proportions, its essence, outside space places and exhibits the constructions in the city. The boundaries, the fences, define the transition between public and private space. The vocabulary of this space must complement that of the architecture. The stratification of the vegetation creates an intermediate height between the building and the pedestrian.

Landscape, the architecture of a land area, is mainly concerned with expressing the limits and the new necessities of a mixed and balanced sharing of the space.

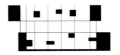 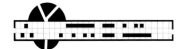  

法国 南特 / 2007

Feltre and Calvaire streets
非尔特尔与卡勒瓦尔街

非尔特尔与卡勒瓦尔街位于南特市中心的历史保护街区，它们构成了南特市区最重要的商业中心。

这个项目将两个普通的街道改造成一个每天有超过300辆公共汽车经过的步行平台。在这个大城市的市中心，机动车的消失、公交车换乘平台和商店送货空间的重新组织，给步行者和树木留出了空间。

这两条街与邻近其他街道和广场之间的连接也都被重新处理，圣尼古拉教堂祭坛的圆形后廊也被改造成大宽度、小高差的台阶，有利于行人通行。

铺着黄色花岗岩的地面区块上设置了一些与公共交通或商业相关的舒适设施。高大的竹丛和单棵的枫树则为城市建立吸引力，并改善了这条街道在南特人心目中的形象。

Situated in a protected historic sector, Feltre and Calvaire streets form the busiest shopping area in the centre of Nantes.

The project has transformed two ordinary streets into a pedestrian zone crossed by more than 300 buses a day. The disappearance of cars, and the organisation of the bus interchange and deliveries, have provided space for pedestrians and trees in a wide, urban courtyard.

The links with the neighbouring streets and the squares have been reworked. In particular the apse of Saint Nicolas's church has been transformed into a series of low-rising, wide steps to facilitate the path of pedestrians.

Blocks of yellow granite house the urban amenities linked to public transport and the shops. Giant bamboo plants and stand-alone sweetgum trees create an attractive urban atmosphere that has raised the image of the street in the minds of Nantes's inhabitants.

左页图：街道轴线和由小广场围塑成的袋状花园。
上图：施工中与完成后的街道。

Opposite page: The street axis and the pocket gardens created on the small squares.
Above: The street during the construction work and after it was completed.

左页图：花岗岩与沥青，经济节约的材料搭配。
上图：小广场与花岗岩地面区块。

Opposite page: Details of the granite and the asphalte, economical materials.
Above: The small squares and the granite blocks.

左页图：道路铺面完全由花岗岩板与沥青两种材料交错搭配。
上图：街道夜间景致。

Opposite page: The paving is all in granite slabs and asphalte.
Above: The nighttime atmosphere in the street.

法国 大莫特 / 自 2011 起

Europe avenue
欧 洲 大 道

被列为法国国家建筑遗产的大莫特市,其发展延续了建筑师让·巴拉度尔作品的一贯风格。它原先是以夏季海水浴为主要活动的城市,随后逐渐发展成为了无季节性限制的城市,其空间的共享性和建筑的高品质是带动城市更新的最重要因素。

欧洲大道被改造成一个朝向大海延伸的带状广场,随着季节的变化,它将其上的三个可依用途而调节的小广场以不同方式展现出来。一个统一的视觉设计规范将大道上的露天座和商店协调起来,其中的城市小品还特别被重新设计过。

La Grande Motte was created in the 1960s by the architect Jean Balladur. Now listed as French architectural heritage, the town aims to continue its urban development in the spirit of the original plan. Designed as a summer beach resort, it became a permanently inhabited town where the sharing of the space and architectural quality are the driving forces for renewal.

The avenue has been transformed into a long plateau that crosses town, ending in the sea. Evolving through the seasons, it brings into play three small squares that can be adapted to different uses. A graphic charter harmonises the terraces and businesses along the avenue, where the urban furniture has been specially redesigned.

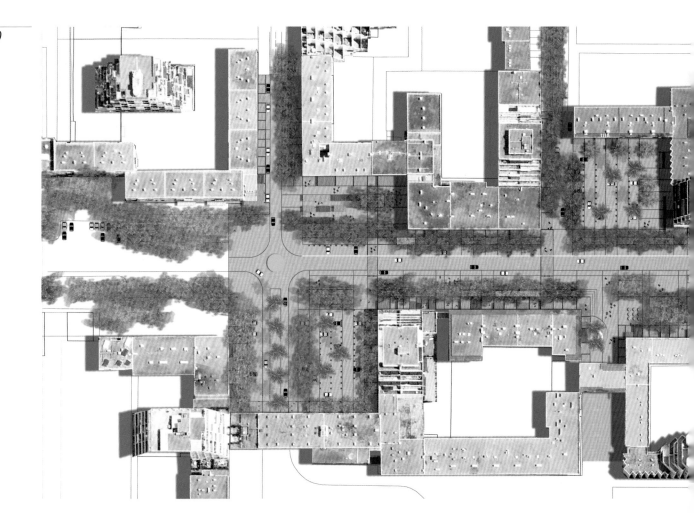

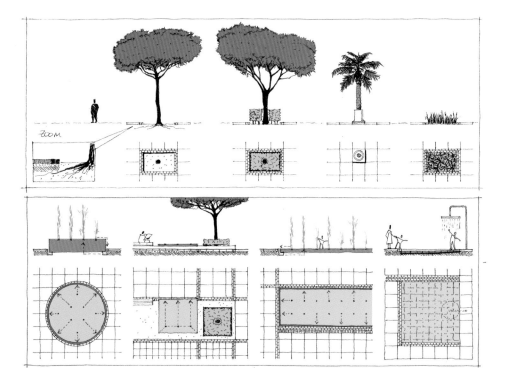

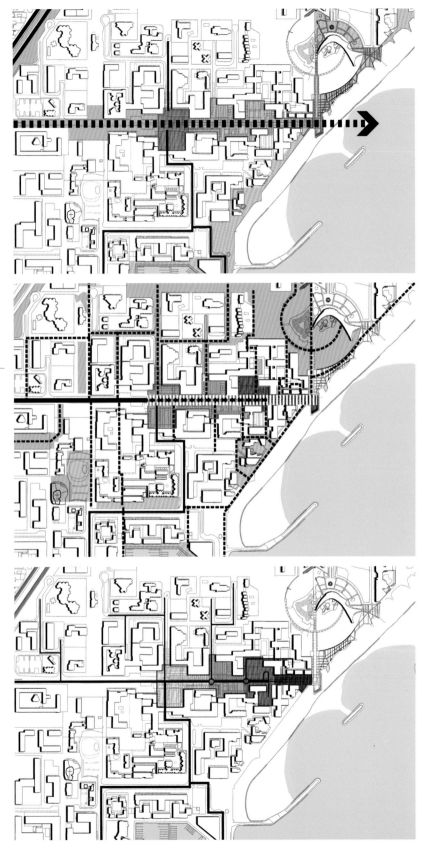

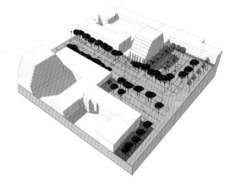

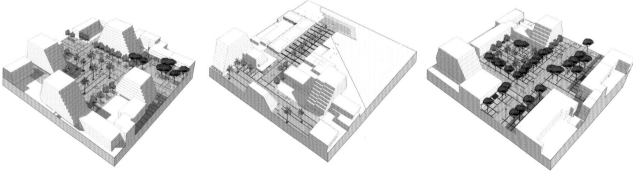

左页图：方案策略。
上图：街道转型为供人分享的广阔广场。

Opposite page: The project strategies.
Above: A street transformed into a shared esplanade.

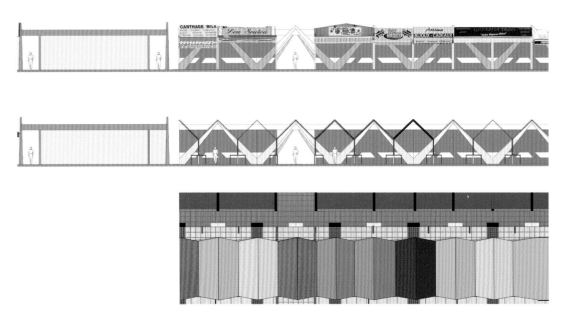

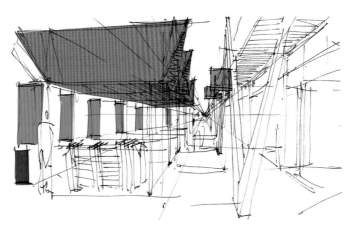
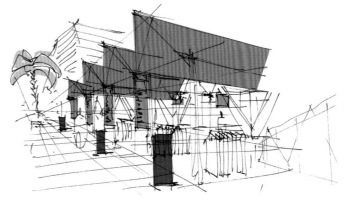

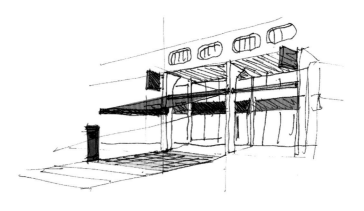

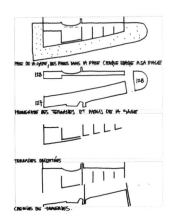 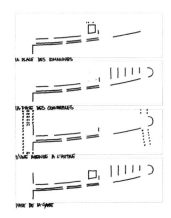 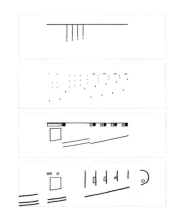

法国 谢西与榭里 / 2002

Ariane square
阿里亚娜广场

在城市本身尚未存在之前就构思它的广场，这是新生城市的城市化节奏所迫使出现的设计挑战。阿里亚娜广场位于马恩河谷，紧邻欧洲谷商业中心和巴黎大区快线RER车站，它为大量的游客敞开了一个通往巴黎迪斯尼乐园设施的入口大门。

除此之外，阿里亚娜广场也必须在城市中心尺度上扮演一个组织性公共空间的角色，将多种功能结合在一起（巴黎大区快线RER车站、商业中心、餐馆的露天座、市立音乐学院、办公楼等）。

广场由三个类型的空间组成：
– 一个种植着排列式植物的大空地，将人们的视线引至餐馆的露天座。
– 一个大广场，连接了大巴黎快线RER车站和一个出现于城市立面上的独特建筑。
– 一个略微低于大广场的花园（位于地下停车场的上方），为附近的居民提供了一个宁静悠闲的空间。

地势高差的处理是巧妙构成广场各个空间的一个重要元素。被修剪成帷幕的椴树本是从当地传统景观语言中汲取的灵感，但这里却以一种不规则的方式种植，形成了历史和现代之间的过渡。

The speed of urbanisation in new towns is such that you sometimes have to imagine a square before the town itself even exists. Located in Marne-la-Vallée, and forming a continuation of the Val d'Europe shopping centre and its RER (high-speed suburban rail) station, Ariane square forms a gateway to the infrastructure of Disneyland Paris for the many tourists who visit.

Beyond this, the square has to fulfil its role as a structuring public space for the town centre. It brings together several uses (RER station, shopping centre, restaurant terraces, the town's music school, offices...).

It is planned around three space typologies:
- an esplanade planted with rows of trees that frame the views towards the restaurant terraces,
- a forecourt forming the link between the RER station entrance and an unusual building in the urban facade,
- a pavement garden (built over an underground carpark) slightly below the forecourt, which offers a quiet space accessible to the nearby residential neighbourhood.

The levelling is a subtle structuring element of the square. The use of pollarded linden trees as a curtain is inspired by the traditional vocabulary of the local square but, planted in an unusual way, they create a transition between history and modernity.

法国 安赞扎克–洛克瑞斯特 / 2007

Lochrist town centre
洛克瑞斯特市中心

安赞扎克–洛克瑞斯特城乡区（居民5800人），坐落在布拉维河畔，位于洛里昂市东北部二十几公里处，在历史上是围绕着三个不同功能的村镇组建的。
– 彭盖斯腾，一个独立的乡间小村庄，特来梅兰森林将它与其他城乡区分离开来。
– 安赞扎克，位于一片农耕土地的中心，是区政府所在地。
– 洛克瑞斯特，是一个工业城镇，从1860年起随着艾乃邦市（布列塔尼铸铁业的中心）的铁匠作坊在布拉维河畔设立而兴起。这里聚集了主要的商业和服务设施。

1960年末众多铁匠作坊的先后关闭使这里逐渐成为一个大型的荒地，但这却是一块特别的滨水地段，拥有着岛屿、水坝和堤岸系统。

洛克瑞斯特公共空间的整治构成了其城市改造的构架：
– 通过洛克瑞斯特的整治来增加市中心密度，使其成为整体城乡区域的中心。
– 在洛里昂城乡群居区域当中加强洛克瑞斯特的区域特征，使它成为一个位于城市和田野之间的"乡村城镇"或者"自然城镇"。
– 持续现有商业并发展新商业。
– 发展混合的和多样化的居住形式。
– 促进柔性交通(步行、自行车、滑轮等)的发展。
– 通过建立布拉维河的防洪安全装置来保障城市安全。

On the banks of the Blavet, around 20 km northeast of Lorient, the commune of Inzinzac-Lochrist (population 5,800) has historically been organised around three settlements with distinct urban functions:
- Penquesten, an autonomous rural village separated from the rest of the conurbation by Tremelin forest
- Inzinzac, surrounded by agricultural land, where the Town Hall is situated
- Lochrist, a workers' town that developed from 1860 with the arrival on the banks of the Blavet of the Hennebont ironworks (the mainstay of the Breton iron and steel industry); the main shops and services are gathered here.

The closure of the ironworks at the end of the 1960s left a large brownfield site, but above all an exceptional riverside site with its system of island, dam and quays.

The regeneration of Lochrist's public spaces forms the framework for a project across the commune that aims to:
- intensify the town centre by developing Lochrist in a way that strengthens it as the heart of the conurbation,
- identify Lochrist within the greater Lorient area as "rurban" and "naturban" – between town and countryside,
- perpetuate and develop the shopping facilities,
- develop mixed and diversified housing,
- encourage soft modes of transport,
- improve safety by putting in place an anti-flood scheme for the Blavet.

经过将近十年的规划工作以及业主的全力推动,这个在城镇本身尺度上显得雄心勃勃的改造方案通过加强历史特色的方式完全改变了城市的形象。

这个方案借助以下方式来组建城市空间:
- 沿着水边开辟一条小路,并在河岸之间建立一条桥道。
- 整治不同的广场,使它们成为一个连贯的系统并重新组织停车场。
- 在布拉维河岸建立河畔花园或者船坞式花园,并重新开辟一条昔日存在的水渠。
- 重新改造市中心的交通干线。

After almost 10 years' work under the exceptional impetus of the local authority, the project, ambitious for a settlement of this size, has allowed us to transform the image of the town while making the most of its unique history.

The public spaces have been strengthened by:
- the creation of a riverside path and a footbridge over the quays,
- the improvement of different squares in the context of a network, and the organisation of parking,
- the creation of riverbank- or dock- gardens on the Blavet, and the uncovering of a stream,
- the improvement of the main artery through the town centre.

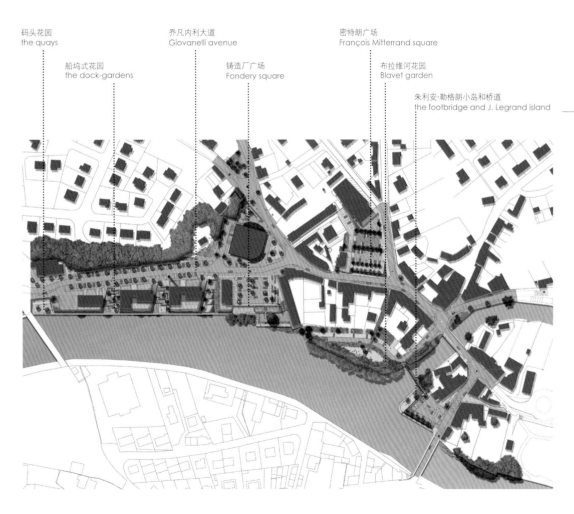

左页图:铸造厂昔日景象和重新整治后的布拉维河河岸。
上图:循着铸造厂旧迹在市中心建造出新的公共空间。

Opposite page: The foundry at the height of production, and the Blavet shore after the regeneration work.
Above: The public spaces of the town centre, evoking the lines of the old foundry.

层层下降的"船坞式花园"将城市与由往昔工业港岸改造成的散步道连接起来。(建筑设计:DDL Architectes)

The public "dock-gardens" link the town to the walk laid out on the old industrial quay. (Buildings: DDL Architectes)

密特朗广场成为不同高度地面与路径之间的连接枢纽。

François Mitterrand square acts as a hinge for the different ground levels and connecting roads.

34

左页图：布拉维河花园与朱利安·勒格朗小岛。
上图：铸铁广场与布拉维河岸，重新整治的空间仍然刻意使用工业语汇来塑造氛围。（建筑改造：David Cras Architecte）

Opposite page: Blavet garden and J. Legrand island.
Above: Foundry square and the Blavet quay laid out with an industrial vocabulary. (Rehabilitated building: David Cras Architecte)

法国 巴黎 / 自2007起

Austerlitz quay
奥斯特利兹堤岸

巴黎的塞纳河岸是被列入联合国教科文组织名录中的世界文化保护遗产，其中广阔且承载着建筑物的奥斯特利兹岸地，具有特殊的形态，并在巴黎东部的城市规划中占有重要地位。快速道路的取消使得河岸得以迎接新的功能，不仅结合了城市和港口的活动、完善了塞纳河沿岸步行道的连续性，也建立了附近街区和塞纳河自然空间之间的新关系。

河岸被改造成为一个以砂岩石块铺地的宽阔广场，形成了一个"缝隙"园地。植物在石块的间隙中生长，适度地提供了有利于生物多样性发展的环境；地面高低差的整合创造出任何人皆可抵达的无障碍环境，同时不影响塞纳河涨潮时的水流；不同地面材料的使用则区分出了个别空间的城市活动功能。

The wide and built-up Austerlitz quayside, which, with other Parisian quays enjoys UNESCO World Heritage status, has an unusual form and poses an unusual challenge in the urban project of eastern Paris.

With the traffic circulation removed, it was possible to plan a programme that unites the activities of the city and the port, extends the riverside walk and creates a new relationship between the neighbourhoods beside it and the Seine.

The quay is laid out as a broad esplanade paved in sandstone, forming a garden of interstices where the planted spaces between the paving stones are, in their own small way, conducive to biodiversity. The levelling makes it accessible to all without hindering the flow of the Seine at high water, and the paving materials have been chosen to prioritise the way the city will use the space.

上图：塞纳河岸的整治，如何让人们重新使用这些巴黎最大的公共空间？
右页图：从废弃的汽车道路转化为新的方案，施工中的河岸。

Above: The quays of the Seine, or how to reappropriate Paris's largest public spaces.
Opposite page: From a brownfield formerly used as roads to the project in progress.

法国 波亚克 / 自2010起
Gironde quays
纪龙德河堤岸

波亚克因其优质的葡萄酒闻名于世，然而，城市工业的消失却使其逐渐黯然失色。

长达1.2公里的纪龙德河畔是此城镇最重要的公共空间，提供了人们广大开阔的地理景观。这些堤岸构成了波亚克城市改造方案的基础。

纪龙德河堤岸采用石灰岩作为立面材质，几个个别带着自然或城市氛围的散步空间互相交接延续。在城镇的中心轴线上，一个紧邻着拉法耶特广场、由河口灌溉的花园提供了城镇居民新的社交空间。

Internationally renowned for its great wines, Pauillac is also a town hard-hit by the disappearance of its industry.

Pauillac's quays, 1.2 km long, are the main public space of the commune. Their nobility comes from their wide landscapes looking out to the geography beyond. They form the basis for a project for improving the town and giving it back its dignity.

From the limestone facades to the shores of the Gironde, several walks succeed each other through built or natural environments. As a continuation of Lafayette esplanade, on the main axis from the town centre, a garden irrigated by the estuary offers a new convivial space.

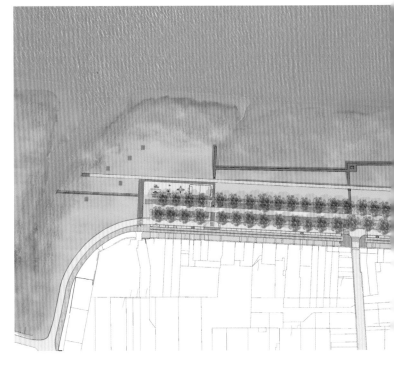

受保护的自然河畔广阔的范围将方案引导至一个改善河口景观和与河水关系的设计思路。

从建筑立面到河岸的行程由三个纵向的散步空间组成：
— "露台"散步空间重新组织在商业建筑脚下的人行道，沿着住宅建筑设置花园，提高地方建筑的价值。
— "巨树"散步空间将法国梧桐戏剧化地展现在人们眼前，同时重新组织了停车空间。
— "河口"散步空间由放宽的围堤展现，人们可以在一座浮桥上欣赏芦苇地和水面的美景。

The broad stretches of protected natural shoreline guided the project towards a plan that makes the most of the estuary landscapes and the relationship with the water.

Three walks run longitudinally from the facades up to the shore:
- the "Terraces" walk reconfigures the pavement in front of the shops and offers a garden running along the front of the residential buildings, enhancing the local architecture;
- the "Great Trees" walk weaves a setting around the plane trees and integrates the parking;
- the "Estuary" walk runs along the existing protective wall, which has been widened, and onto a pontoon from which you can watch the reedbed and the river.

三个散步空间：平台、林荫道和纪龙河畔。

The three walks: the terraces, the avenue of trees, and the Gironde.

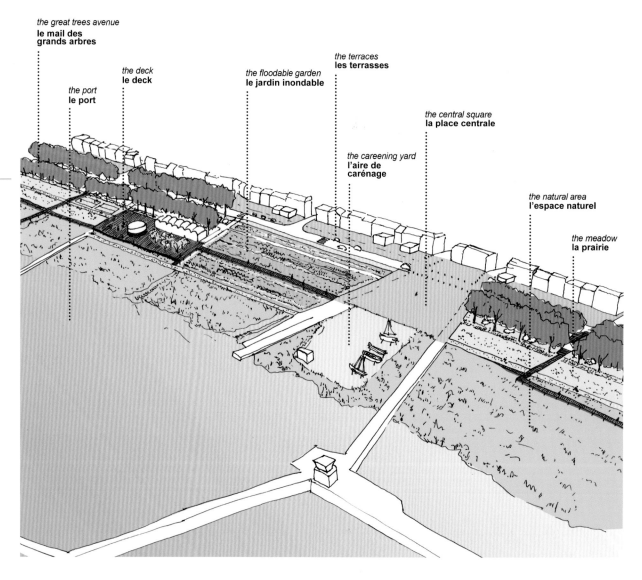

上图：由纪龙河灌溉的花园以及它们与河岸的各种连接。
右页图：从汽车道路转化为舒适街道，重新被人们使用的公共空间。

Above: The garden irrigated by the Gironde and its links to the quays.
Opposite page: From busy road to promenade, the reinvestment of public space.

左页图：纪龙德河畔伸展台般的散步道；堤道小品的创作灵感来自于精致纯熟的木桶工艺。
上图：由目前港边延伸的人造土堤转变为未来可吸收涨潮洪泛的湿地花园。

Opposite page: The landscaped Gironde walk; the furniture is made using traditional cooperage skills.
Above: From the existing dockland terrain to a floodable garden.

法国 CCPHVA 惠桑泽平原 / 2009

Beler gateway
贝雷尔城门

贝雷尔城门项目位于法国和卢森堡边境处的法国摩泽尔省和莫尔特-摩泽尔省，它的对面是规模庞大的贝尔沃项目(将创造20000个就业机会和5000户住宅)，因此，贝雷尔城门项目正处于一个生态城乡区的蓬勃发展中。

这块具有吸引力的开发用地，四周围绕着用于休闲用途和绿色农业的乡村景观，非常适合建造提供卢森堡人使用的经济型住宅、跨国界的城市设施和商业建筑。景观构思在这里成为一个同时兼顾经济、社会和环境的重要元素，将一条车行公路改造成为充满生机的林荫大道，让共同历史成为方案的资源，从而创造出一个两国人民共享的居住盆地。

项目以贝雷尔河谷为中心而展开，此河谷是卢森堡主要河流阿尔泽特河的起源。方案的构思超越了国家边界，围绕着一些重要主题而逐渐成形：可持续发展、两国功能运作的互补、环境的连续性、交通的共享逻辑以及有利双方国家经济的发展策略等。

This site, which straddles the two French departments of Moselle and Meurthe-et-Moselle and faces the ambitious Belval project (20,000 jobs and 5,000 homes), forms part of the development strategy for an eco-conurbation on the France-Luxembourg border.

Organised around rural landscapes dedicated to leisure activities and nearby organic agriculture, these attractive French land reserves offered an opportunity to create affordable housing for Luxembourg, cross-border amenities and opportunities for trade. Transforming a road into a residential boulevard, and drawing on a shared history to form a cross-border population pool, landscape architecture is here an economic, social and environmental project.

The project is built around the Beler valley, the source of Luxembourg's main river, the Alzette. Thinking beyond borders, the approach is guided by the theme of sustainable development, the complementary nature of the two countries' planning strategies, environmental continuity, shared policies on border-crossing, and development strategies that make the most of the strengths and weaknesses of both states' economies.

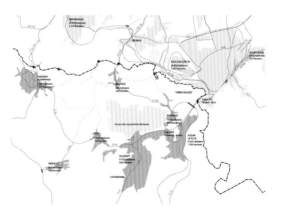

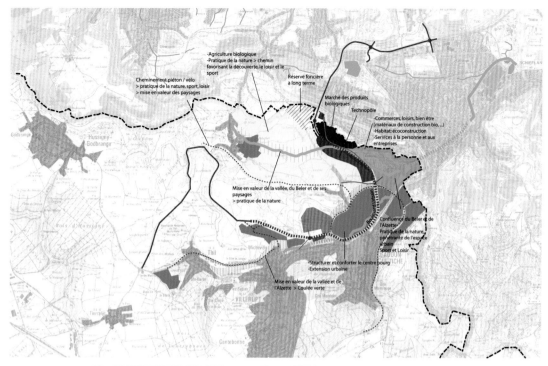

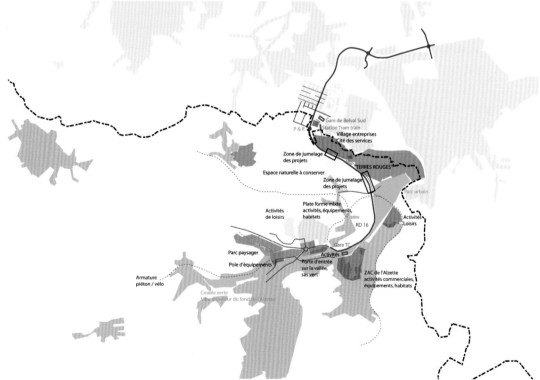

左页图：方案进行前的场地现况；法国和卢森堡边境地区的道路、城市与景观网络分析。
上图：功能分布计划图与进行中项目综合图，以使该方案成为一个协调有致的土地整治项目。

Opposite page: Before work began: the road, urban and landscape networks of a cross-border territory.
Above: Planning map and synthesis of the projects in progress for a coordinated land development scheme.

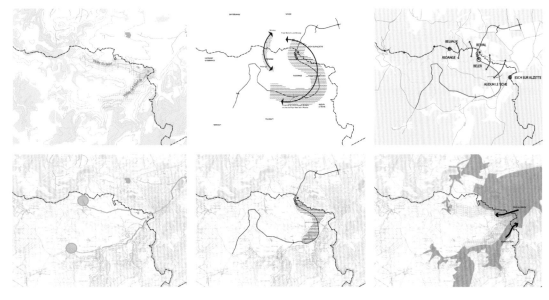

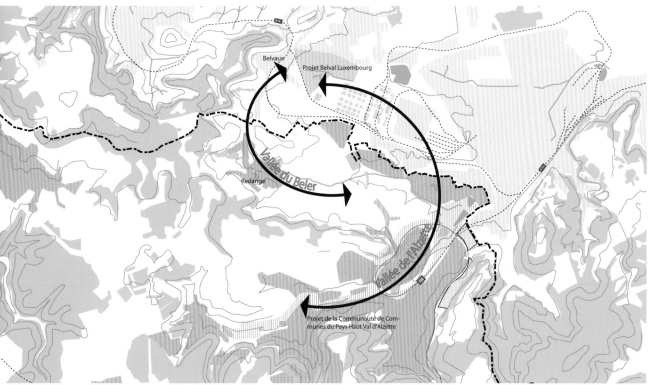

左页图：方案进行前的场地现况；阿尔泽特河与贝雷尔河河谷，法国和卢森堡边境地区的景观与贝尔沃城门（卢森堡）。
上图：空间阅读与土地特征研究，作为互补性策略的计划基础。

Opposite page: Before work began: the Alzette and Beler valley, a cross-border landscape, and the Belval gateway (Luxembourg).
Above: A reading of the space and the strengths of the territory, designed to encourage complementary planning.

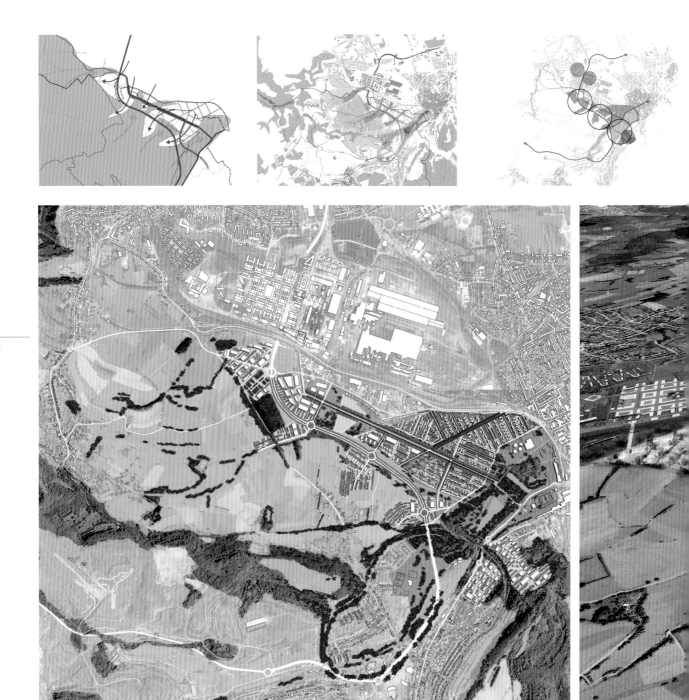

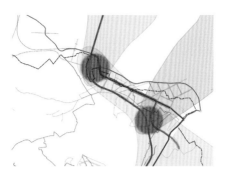  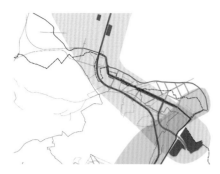

两页上方图：地域尺度的整治构思，含了城市、经济、景观与生态环境等主题。
两页下方图：方案整体配置图与透视效果图。

Opposite page and this page, top: The development of the territory by theme, showing its urban, economic, landscape and environmental impacts.
Above: Perspective ground plan of the project.

法国 小罗塞尔 / 2010
Saint-Charles pit
圣夏尔矿井

本项目所在城市小罗塞尔位于德国边境的一个早先以矿业和钢铁业为经济活动的地区，在这些工业产业停止经营以后，城市试着寻找着它经济发展的第二春。

小罗塞尔的城市形态来源于19世纪末开始的地下矿产开采：地下矿业设施严重破坏了地貌，工人街区的发展则体现了矿井逐渐开发的不同年代，一些路径让人回想起当年工人的足迹，而所谓的自然空间也只不过是那些被弃置不顾的未开采空间。城市的历史和形态与矿业的历史紧密结合、不可分离。

如今，无论从空间角度或是社会角度，这个城市都无以为继：住宅逐渐损毁，商店关门，服务设施不足，甚至连市中心都难以辨认。

我们所发展出来的城市发展指导计划借助边境城市的活力和工业旅游的规划来使这块四分五裂的土地产生和谐并重新获得生命力。

新城镇的城市方案建立在已经消失的矿业的遗迹之上：矿井的井架成为连接市中心和矿工生活区的地标，在昔日的马厩建立了文化中心，在工厂主轴线上开设了一条连通各街区的路，原先的铁路被改造成自行车道，过滤池和周边森林也被整治成城市自然公园。

Located on the German border in a mining and ironworking region, Petite Rosselle is looking for a new lease of life after the demise of its industry.

The town was shaped by the mining that went on here from the end of the 19th century: its topography has been uprooted by the mining infrastructure, the concentrations of workers' housing reveal the chronology of the successive pit openings, the paths retrace the workers' routes, and the "natural" spaces are the neglected areas that could not be exploited as mines. The history of the town and its urban form are indissociable and interlinked with that of the mine.

Today the commune is in limbo, from a spatial as well as a social point of view: the housing is run-down, shops are closing, there is a lack of services and the town centre is hardly recognisable as such.

The guide plan that has been developed gives coherence to this fragmented territory while revitalising it through cross-border dynamism and industrial tourism.

The urban project for the new town is based on the lines of the former mine. The mine headframe will become a reference point around which the centre and the mining estates revolve; the cultural centre will be in the former stables; a street that opens up the neighbourhoods is placed on the main axis of the factory; the railway lines become a cycle path; the settling ponds and the surrounding forest will be developed into a natural urban park.

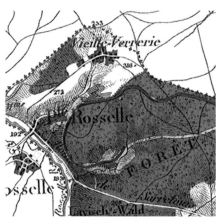

Cassini map 1750 / 卡西尼地图　　　　1880　　　　1930

53

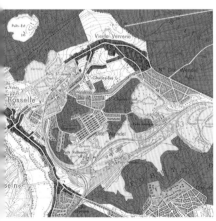
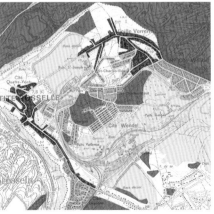
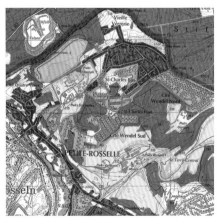

1955 1975 2009

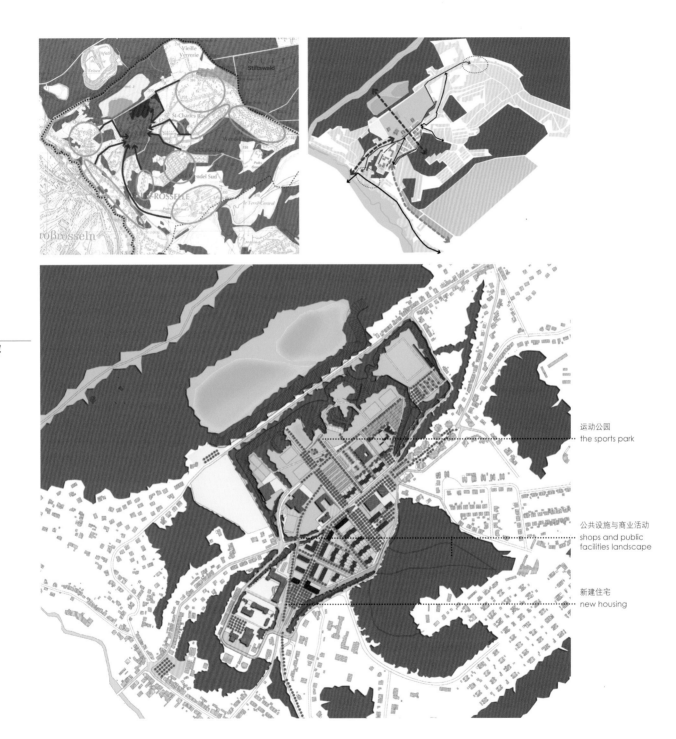

运动公园
the sports park

公共设施与商业活动
shops and public facilities landscape

新建住宅
new housing

上图：新街区的空间组织原则。
右页图：在工厂城市的时期之后，城市完全取代了工厂。

Above: Organisational principles of the new neighbourhood.
Opposite page: After the factory town, the town replaces the factory.

位于市中心的工厂被新街区所取代，新街区则围绕着体育和文化活动以及新住宅的建立来组织规划。新住宅在建筑的形态和密度上都从传统工人住宅区中获取了灵感，这些工人住宅如今已成为象征当地历史的建筑遗产。

新街区成为了边境城乡居住区自然公园的中心，也是和众多景点保持联系的旅游中心，例如：法国最大矿业基地卡胡-万得尔的几个矿业博物馆，以及现存完整的德国冶炼大厂沃尔克林格耐斯特工厂。本项目区域预计在2020年完成140户新住宅，通过其历史和景观而焕发出新生命。

The factory at the centre of the town has been replaced by a new neighbourhood organised around sports and cultural activities. The new housing is inspired by the form and density of the workers' estates.

The new neighbourhood will become the centre of the Cross-Border Conurbation Natural Park, a tourist centre linked to the museums of Carreau Wendel Mine, the largest mining site in France, and the Völklingenest factory, a German steelworks that has remained intact. With 140 new homes on the horizon in 2020, this site is regenerating thanks to his history and its landscape.

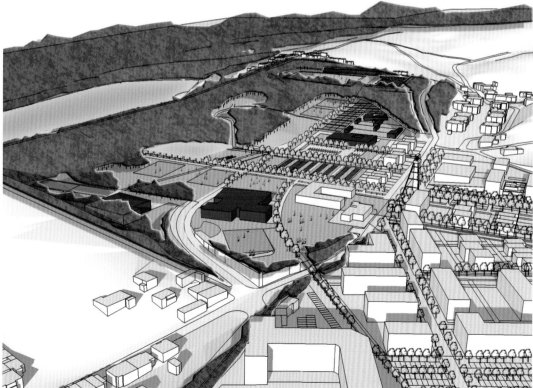

万德乌市的鸬鹚街区。 The Cormorans neighbourhood in Vandœuvre.

整体的方案是共创的方案
a project as a whole, a project together

人们以住宅为中心展开一系列活动：停车、处理垃圾、游戏、种植植物、散步、与人会面、赋予这个"自己的家"一种形象、保证自身的安全，并且限定生活空间的边界。室外空间是改善城市实践的潜力空间，它该如何面对生物多样性、水处理，软性交通方式、垃圾的合理管理、停车位的共享、城市声效和热效、阳光和阴影、共享花园和邻居晚会？室外空间在城市和建筑建造中经常被遗忘并且疏于管理，最多成为装饰性空间，它们应该在功能使用的基础上被重新进行整治。

我们主张对空间使用的掌握，以使得人们对社会面貌的需求和理解更加细致化。

居民协商和居民参与是一种有益的实践方式，可以丰富功能计划、带来适合不同人群的解决方案，并且改善因方案而产生变化的接受条件。

我们只是方案中一个部分的参与者，那就是构思和实施，这是项目中必不可缺的工作，是对业主需求直接相关的表达。居民的参与，政治意愿的永久性也是一个成功的共享方案的不可缺少的组成部分。只有人们共同创造，才能达成一个整体性的方案。

Around our homes we park our cars, we manage our refuse, we play, we garden, we walk, we meet each other, we give an image to "our house", we make ourselves safe and we mark out the limits of our living space. Outside space has the potential for improving urban practices. What place is given to biodiversity, to the alternative management of water, to soft means of transport, to the well-considered management of our refuse, to pooled parking, to controlling noise and heat pollution, to light and shade, to community gardens and neighbourhood get-togethers? Often neglected, at best decorative, outside space is all too often forgotten in urban and architectural production. We must invest in it once again with a project based on use.

The use of the city is in constant evolution; new eco-citizen practices are emerging, and the space is used differently according to the weather and social origins. But how can we think sustainably about the city of tomorrow if these evolutions are not the departure point for our thinking on the architecture of the outdoors? How can town-planning experts be heard if the expertise of daily life is not recognised?

We are in favour of bringing in a means of measuring and studying usage that would refine the understanding and the demands of each social landscape. The consultation with and participation of the inhabitants are both means of enriching the programme, of bringing solutions adapted to different populations, of improving the conditions that will lead people to accept the changes linked to any project.

We only act on one part of the project, that of its design and implementation. The work of setting up the programme, the pertinent expression of the commission, the participation of the inhabitants and a consistent political will are also indispensable contributions to the success of a shared project. Any project that is whole must be done together.

法国 万德乌-雷-南锡 / 自1998起

Vandœuvre town centre
万德乌市中心

万德乌-雷-南锡是法国莫尔特和莫泽尔地区的第二大城市（居民3.1万人），在1960年代从乡村转变成优先城市化地区，大量建造了交错排列在道路网络之中的塔楼和板楼建筑。

在这最初的发展之后，万德乌从1990年代起开始经历人口的衰退、住宅形式的不适应、公共空间的逐渐陈旧、居民的老年化和贫困化等问题。

万德乌市城市改造项目于是从1998年开始展开，由宇比库斯事务所负责进行方案构思。

Vandœuvre-lès-Nancy (population 31,000), the second largest town in the Meurthe-et-Moselle department, was a rural Lorraine village that became a Priority Development Zone (Zone d'Urbanisation Prioritaire – ZUP) in the 1960s, when its towers and horizontal blocks were interwoven with the road infrastructure.

After this initial development, the town has gone into decline since the 1990s from a downslide in its demographic caused by the unsuitability of its housing, the obsolescence of its public spaces and the increasing poverty and ageing of its population.

Urbicus has been working on an urban restructuring project in the town of Vandœuvre since 1998.

方案进行之前的基地状况。

The site before work began.

这个城市改造项目位于一个160公顷的城市敏感区域中，其中包含了三个街区、市政府周边地块和国民商业中心。

国民街区是万德乌市一个处境艰难的商业中心，同时也是万德乌市的市中心。它被欧洲大道（车流量30000辆/天）隔离，是整个改造项目的重点，目的在于重新建立万德乌的商业吸引力。

This urban renewal project concerns three neighbourhoods as well as the area around the Town Hall and the Nations centre. It is at the heart of a deprived urban area covering 160 hectares.

The Nations is both Vandœuvre's ailing shopping centre and its town centre. This central point, which is cut off by Boulevard de l'Europe (30,000 vehicle a day), is the focus of a redevelopment project that aims to make Vandœuvre an attractive shopping centre again.

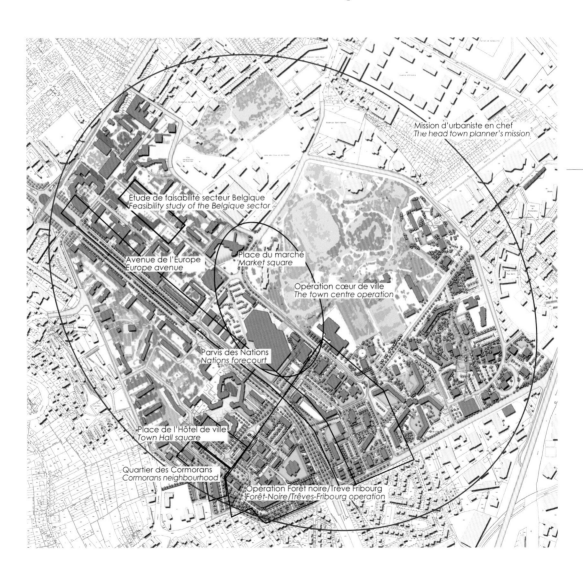

研究区段的整体平面配置图。

Overall plan of the sectors studied.

将城市分割成两部分的凯尔桥被列入拆除计划，这为加强城市密度提供了新的可能性。

特来乌-弗赖堡和黑森林等街区是城市改造项目的重点地区，改造内容包括：街区边界、功能、多样化住宅的重整以及城市特色更新等。鸬鹚街区主要由一个原来长200米的住宅板楼组成，这个板楼被划分成几个段落，以使得街区能够向重新整治过的市政府广场开放。

这项城市重整的工作也以万德乌的山丘景观特质为基础，以促进和加强其公园城市的品质与特色。

The Kehl bridge, which cuts the town in two, will eventually be demolished, opening up new possibilities for urban housing.

Trèves-Fribourg and Forêt-Noire neighbourhoods will be restructured to modify their peripheries, open up new uses, rebuild diversified housing and reforge the urban identity. The Cormorans neighbourhood is essentially made up of a horizontal housing block 200 metres long that has been split up to open up the neighbourhood onto the redeveloped Town Hall square.

This restructuring work draws on the attractive landscapes of the Vandœuvre hillsides to enhance and strengthen its identity as a garden city.

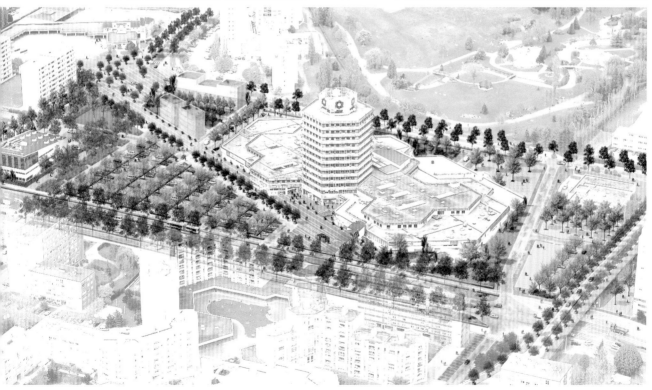

左页图与上图：方案进行前后景象对照。以植物作为城市街区重整的主要元素。

Opposite page and above: The project before and after. Plants as the vocabulary of the urban restructuring.

鸬鹚街区，2004整治之后的街区景象。

The Cormorans neighbourhood after its completion in 2004.

法国 维特里-勒-弗朗索瓦 / 自2004起

Vitry suburbs
维 特 里 郊 区

维特里-勒-弗朗索瓦是马恩河省的第四大城市,有1.5万人口,弗朗索瓦一世于16世纪在此兴建了城墙。它位于运河、铁路和公路的交汇处,这些交通线路曾经促进了城市最初的发展,但如今已经不能保证城市未来发展的需要。以大型居住区模式建设的社会住宅逐渐变得贫困、封闭、遭受废弃和贬值。

本改造方案致力于城市空间的重整,将住宅区与火车站和市中心重新连接起来,让城市朝向水面开放,并开发新的住宅区。方案也在城市尺度中建立起一个清晰的公共空间网络,并在其中开辟了一个新的公共花园、重新整治了火车站的前广场,这些措施减轻了由于车站汇集大量铁道所产生的城市断隔。

在2005年赢得项目的规划研究之后,事务所接着投入了预计持续12年的设计和执行工作,以重新建立城中心和郊区村镇之间的联系。

Vitry-le-François (population 15,000), the fourth largest town in the Marne department, was fortified by Francis I in the 16th century. It is situated at a crossroads of canals, railways and roads that encouraged its initial development but are now insufficient to ensure its future. Its neighbourhoods of public housing, based on large blocks, are impoverished, isolated, obsolete and run down.

The project aims to rehabilitate these areas, connect the neighbourhoods to the town centre and the station, open them up to the riverbanks and build new housing. The sense of disconnection caused by the mass of railways lines is reduced by the creation of a public garden and the rehabilitation of the station square, forming part of a network of public spaces across the town.

After a definition study submitted in 2005, the town decided to commit to 12 years of work with unwavering determination, in order to weave new links between the town centre and the suburbs.

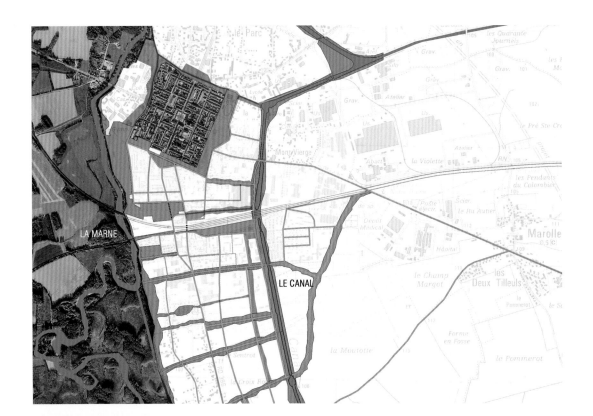

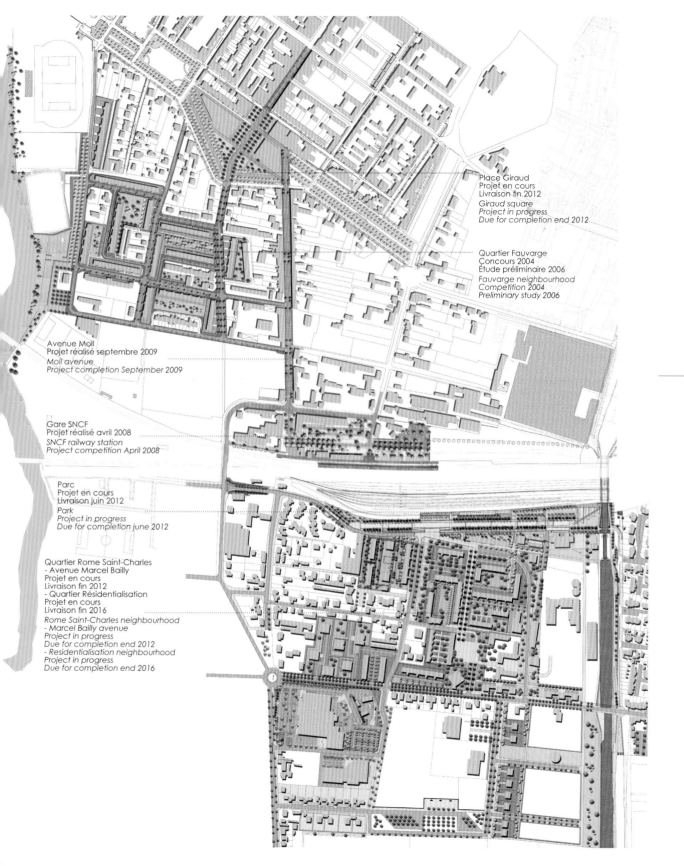

2011-04

2011-05

2012-03

2011-04

2011-05

2012-03

66

2011-07

2011-10

2012-03

左页图：方案随着时间的推进而逐渐实现，让人看到基地的明显转变。
上图：中央绿毯。

Opposite page: Transformation of the site and "construction" of the project over several months.
Above: The central lawn.

为了迎接高速铁路的到来，火车站被改造一新。重整后的车站广场成为本街区与城市中心的连接枢纽。

The station square, redeveloped as a new articulation point between the neighbourhood and the town centre for the arrival of the high-speed train line.

法国 图勒 / 自2006起
Croix de Metz neighbourhood
梅兹十字街区

梅兹十字街区的总体规划是图勒市自2006年以来所进行的城市更新项目之一。其板楼式住宅建筑的拆除使人们对该街区的公共空间有了新的构思，同时也确认了将其塑造为具有高大树木的低密度"公园式街区"的想法。

街区的东西两个部分分布着众多的广场和小径，方案利用位于中央的勒克大道将它们整合起来。

每栋住宅建筑都被重新组织以提高不同空间的可识别性：建筑边界、停车场和花园。一个步行小径系统被融入了住宅区的空间架构网络中，以便于加强小区之间的联系。

The whole of the Croix de Metz neighbourhood has been the subject of an urban renewal project since 2006. The demolition of horizontal housing blocks allowed us to rethink the neighbourhood's public spaces and to strengthen its character as a "park neighbourhood" with law housing density and planted with large trees.

Leuques avenue allowed us to unify the east and west part of the neighbourhood, run through with small squares and lanes.

Each housing block has been readdressed and reorganised in order promote the readability of the different spaces: the threshold of the building, parking and garden. A network of footpaths has been placed inside the housing grid, encouraging the neighbourhood's residents to meet each other.

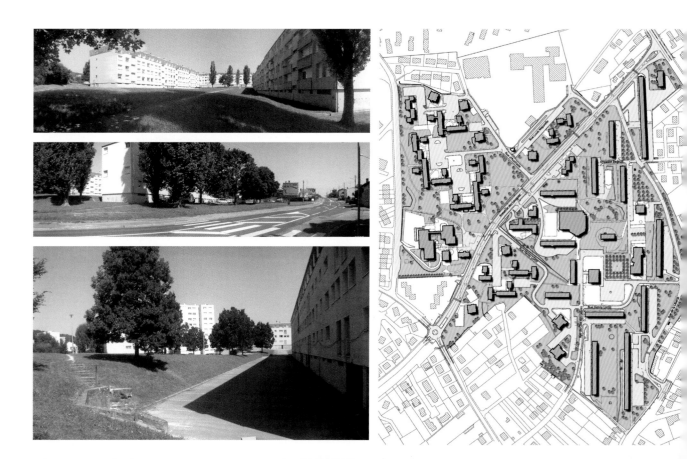

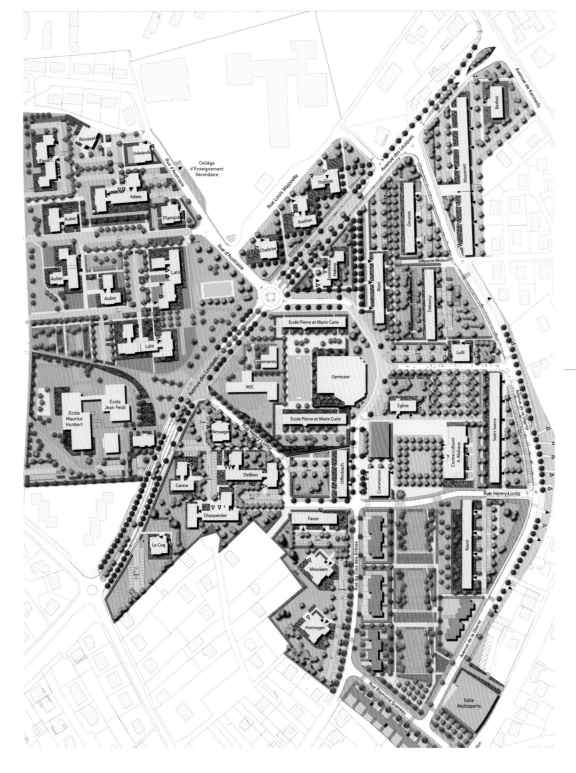

左页图：方案进行前的基地状况，一个封闭而穷困的街区。
上图：街区改造方案的整体配置图，通过景观设计与新建筑的建造来重新整治街区。

Opposite page: Before work began, an isolated and run-down neighbourhood.
Above: Site plan of the neighbourhood restructured by the landscape and rebuilding.

上图：勒克大道，从原先的汽车道路转变为城市林荫道。
右页图：让·费德广场，丘陵景观也成为此公共空间的重要元素之一。

Above: Leuques avenue, a road transformed into a boulevard.
Opposite page: Jean Feidt square – the hillside landscape becomes part of the public space.

左页图：一个花园街区的住宅与小径。
上图：公园施工前和施工中景象，展现出它与街区本身以及周围大型景观的关系。

Opposite page: The apartment blocks and footpaths of a garden neighbourhood.
Above: The park before work began and during construction, revealing the relationship to the neighbourhood and the larger landscape.

"自然城市性"   the "naturbanity"

城市通常从中心朝外扩张或者一块块地吞噬着乡村空间，而乡村总是在城市的边缘承受着来自城市的压力。大批的乡村人口移居城市以后，轮到城市来对乡村进行消耗。城市土地的价格高昂，人们对城市的喜爱冷却了下来，乡村因而变得更具有吸引力。如今广泛使用的"城郊结合"系统，却过度地侵占了农业空间，也造成基础设施匮乏不足，并引发交通问题。

在现今多样化、全球化的社会里，由于各式各样的网络建构与交错，一切空间，甚至自然空间都产生城市化的趋势。自然空间这个旅游业的媒介，已经变成了支援经济活动的筹码。观光人潮过多的自然空间因此需要受到保护，并重现大自然的景观。

城市的蜂蜜比乡村的蜂蜜质量更佳。有机农业在城市中寻找土地，以便发展靠近消费者的市场，而农药则污染了我们的乡村和花园。

Concentrically or in patches, the city absorbs rural land. The countryside, which is always at the edges of cities, suffers from urban pressure. After a human exodus from the countryside to the city, it's the city that consumes the countryside. The lower price of land, and disenchantment with the city, makes the countryside more attractive. "Rurbanity" is a dominant system today which brings with it an over-consumption of agricultural space, a lack of facilities, and transport problems.

In our global society, thanks (or no thanks) to networks of all kinds, everything is becoming urban – even nature. The natural space itself attracts tourism and becomes an economic support. When it is over-used, natural space needs to be returned to nature and protected.

The honey from beehives in cities is better quality than the honey from the countryside. Organic agriculture seeks land in cities in order to develop a local market; meanwhile, pesticides pollute our countryside and our gardens.

这个新世纪的挑战是将城市周边的地带并入城市、控制城市的扩张，并保护乡村，让农产品不再受到污染，同时维护自然空间中各种生物的生态平衡，以重新提升自然空间的价值。"自然城市性"便是与这个城市运动相符合的景观。

享有土地所有权是法国大革命的成果，享有景观权利则是通过对环境进行改革来增加城市的密度，并提高城市活动的强度。

在所有不同形式的花园住宅区中，建筑形态决定了城市形态，而我们建议考虑一种由空间来界定实体的景观城市。

通过景观来提高城市活动的强度是我们对这个课题的建议。发展"城市中的自然环境"、建立以共享花园与城市农业为基调的新型"城市自然公园"、由绿化来建构崭新的城市全景，这些都是经过深入探讨、以景观来限制城市扩张的解决方式。

The challenge of this nascent century is thus to bring the urban back to the cities, to limit urban spread, to preserve the countryside for non-polluting agriculture, and to restore natural space under the flag of preserving biodiversity. "Naturbanity" is the landscape equivalent of this urban movement.

The right to property was acquired during the French Revolution; a "right to the landscape" would allow a densification of the cities and an urban intensification, while structuring an environmental revolution.

After all the variants of garden cities, where the architectural form has determined the urban form, we are offering the idea of a landscape city where the void determines the solid.

Urban intensification via the landscape is our response to this problematic. The development of "nature in the city", the invention of new "natural urban parks" made up of community gardens and urban agriculture, and the structuring of new urban views through planting are among the solutions developed to circumscribe the city with the landscape.

索恩河畔沙隆的布莱-得旺公园。 Prés-Devant Park in Chalon-sur-Saône.

a "positive city"
"正能源城市"

要求在城市中创造自然是城市规划项目中一个经常被提及的主题，我们应该赋予这个主题意义和内容。城市中的自然景观如果没有进行可持续性的设计，就有可能成为单纯的城市装饰品。以绿化墙或轻轨电车的草坪轨道平台为例，两者都是媒体上广受好评的例子，实际上却都消耗大量的水分，与可持续发展的观念背道而驰。

城市是否能创造出一个积极能源的环境，如同那些能源产量大于耗量的积极能源住宅？要建立一个"正能源城市"，我们必须让组成城市的线条与轴线更加丰富多样。

无论是面向开阔的景观或是私密的空间，人们视线的交错都需要透过虚与实所构成的体量作为引导，借以塑造出近身环境尺度和大区域尺度的空间。

绿色网络和蓝色网络可以产生生态环境的多样性和水的节约，但它们只有在超越土地所有权的界线和个别项目范围的界线、在保持其连续性的情况下才具有实质的意义。

空间的时间性促使我们在构思方案时便必须考虑它的持续发展，包括循序渐进的实施过程，以及可逆转和可适应的能力。将未来的管理模式列入设计考量是一个项目持久的保证，它能将有利的因素融合到项目自身的发展当中。

空间的边界和其精确建立起的渗透程度为各种活动提供了清楚的架构，并保证了每个空间在城市中的识别性。

在城市形成的过程中尊重历史遗迹与地理形态，能够加强城市自身的特点，避免了缺乏根基的拼凑成形。

我们将城市的这些不同层面合理而有次序地叠加起来，借以构思"正能源城市"和规划生态街区，我们将其称之为"景观城市"。

Nature in the city is a recurrent request in urban projects, which must be given sense and content. Such a landscape risks being merely decorative if it is not designed to be sustainable. Vertical gardens or grassy tram platforms have attracted a lot of positive media coverage, but they are counterproductive in terms of sustainable development because they are great consumers of water.

Can the city positively generate environment in the same way that positive energy homes produce energy? To weave this positive city, we must work on enriching the warp and weft that compose it.

The mesh of viewpoints, from those that survey the landscape as a whole to those who think in terms of privacy, calls for a volumetric strategy of solids and voids that sculpt the space on scales that range from close proximity to that of the land area.

Green and blue grids, which can generate a diversity of environments and economy of water use, only work in a continuity that crosses the boundaries of land ownership and the operational perimeters.

The temporality of space forces us to think of the project in the long term by integrating the progression of its implementation, and a capacity for reversibility and adaptability. Taking the way it will be managed into account guarantees a sustainable project with the genes useful for its own evolution written into it.

The weave of boundaries and the precise noting of its degrees of porosity frame our practices. They ensure a reassuring sense of address situated in the city.

The sedimentation of historical traces and geographical layers in the making of a city strengthens the identity-forming references that avoid rootless collages.

We superimpose these urban layers in a reasoned and localised way, based on an order of priority, to design this positive city, this eco-neighbourhood project that we call a "landscape-city".

法国 安德莱西 / 2009

Seine hillsides
塞纳河畔丘陵地

巴黎地区的居民面临着越来越严峻的住房问题。大巴黎地区所有的城市规划研究都提出了建造新住宅的必要性。安德莱西丘陵上的可建造地块拥有良好的公共交通服务，提供了城市化发展的好条件，但却似乎与保护环境的需求背道而驰。

此地的景观来自于生态环境和地理形势、来自于意义重大的广阔视野和场所特性，也来自于提供生活需求的农业活动和蔬果园艺。它被仔细地研究以便成为城市设计中具有整合空间和提升基地价值之作用的恒常性元素。

此项城市规划将自然空间和农业用地的三分之二面积用于建立一座属于周边城乡区域尺度的自然公园，成为城市化的基调。它融入于一块开发绿色农业的地块之中、在河流和森林之间开发生态走廊、种植用来界定地块边缘的树林、建立与公共交通相连的道路网，并且利用一系列的平台式建筑来取得开阔的视野。

方案还将一条汽车道路改造成行驶轻轨电车的林荫大道，以便建立一个新的城市中心，并与塞纳河沿岸的老城区维持联系。

这个新颖的城市改造经济模式也是一项对可持续发展之景观的保护策略，它同时也避免了土地的荒废和住房的匮乏。

The French have been confronted with en epidemic of poor housing. All the town planning documents for the government's Grand Paris (Greater Paris) initiative talk of the necessity of building housing in the Ile de France region. The land available on the Andrésy hillsides, already served by public transport, represents an opportunity for urban development, but is also confronted with the apparently contradictory need to preserve the environment.

The landscape, made up of ecology and geography, of iconic views and the attachment of identity, of agricultural activity and subsistence gardening, is described in detail in order for it to become a structuring, newly valued and intangible element of the urban project.

The protection of two thirds of the wild and agricultural land through the creation of a Conglomeration Natural Park structures the town planning. It slides in between the plots of nearby organic agriculture, places ecological corridors between the river and the forest, develops the hedgerow vegetation structuring the borders and weaves in pathways to meet public transport, while a system of architecture in terraces opens up the views.

The transformation of a road into a boulevard on which a tramway runs will facilitate the creation of a new town centre linked to the Seine-side quays of the old centre. This new urban economy is a preservation strategy for a sustainable landscape that offers an escape from the problems of land falling into neglect and a shortage of housing.

- 如何建立河谷与丘陵之间的关系？
透过一系列散步道和景观结构来连接塞纳河、河谷坡地和欧提勒丘陵。
- What will be the relationship between the valley and the hillsides?
A network of walks and landscape structures will link the Seine to Hautil via the hillsides.

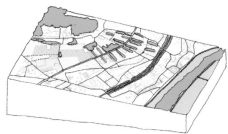

- 如何建立城市与自然空间的关系？
安德莱西的丘陵地成为提供居民使用的广大公园，沉浸在耕植式景观之中（家庭式蔬果园、城郊农业田园……）。
- What will be the relationship between the town and the natural space?
The Andrésy hillsides will become a large inhabited park in a cultivated landscape (vegetable plots, peri-urban agriculture…).

- 应该建立何种城市密度与强度？
建立高密度的住宅地块以使景观能够获得维护。
- What will be the urban density and intensity?
Blocks with high density to preserve the landscape.

- 如何提供街区便利的城市交通？
建立一系列便于通往轻轨车站的路径。
- What access to public transport will the neighbourhood have?
A network of pathways that lead to the station.

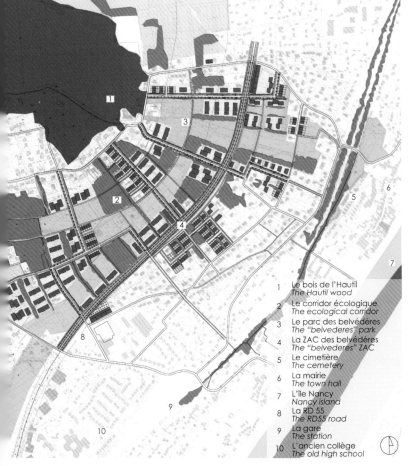

1. Le bois de l'Hautil / The Hautil wood
2. Le corridor écologique / The ecological corridor
3. Le parc des belvédères / The "belvederes" park
4. La ZAC des belvédères / The "belvederes" ZAC
5. Le cimetière / The cemetery
6. La mairie / The town hall
7. L'île Nancy / Nancy island
8. La RD 55 / The RD55 road
9. La gare / The station
10. L'ancien collège / The old high school

法国 蒙泰夫兰 / 自2006起
Mount Évrin park
艾夫兰山公园

位于昔日农业用地上的艾夫兰山公园（面积大约20公顷）是即将建成的新街区的脊柱。公园延伸1.8公里之长，连接了坐落在马恩河畔小山丘上的古老村镇和在高原上刚刚开发的马恩河谷新城的公共服务设施。

公园享有周边乡村特有的田园景观：草地、灌木围篱、小树林以及一个将用于建立雨季储水池的大型湿地。大果园位于住宅区正中央，呈现出非常田园化的氛围。

Situated on old agricultural land, Mount Évrin park (around 20 hectares) forms the backbone of a new neighbourhood yet to be built. Stretching over 1.8km, the park provides a link between the old town, on a hillside in the Marne department, and the infrastructure and services of the new town of Marne-la-Vallée developed on the plateau.

Its atmosphere draws on the rural landscapes of the surrounding countryside: meadows, hedgerows, small woods and a broad wetland that came with the creation of a storm basin. At the heart of the housing, the Grand Orchard offers the most garden-like atmosphere in the park.

一些园圃种植了果树、栗子树和榛子树，提供前来散步的居民摘取果实。另外，种着树的开敞草原延续着金雀花原野而展开，并通向一个新生的小树林。

这个作为自然空间而被使用的公园将随着具有可持续性的多样化管理而不断演化，目的在于提高生物多样性和节约维护费用，因而管理者只在小路的边缘进行规律性的修剪，草原的维护则随着植物的开花季节和游人的数量而灵活调整。水沟和树篱被设计成为生态走廊，种植着简朴的地区性植物。

类似的自然活力也体现在相邻的生态街区的构思上：景观元素潜入了私人住宅地块。它们不仅有助于空间用途的规划，也在公共空间与私人空间之间建立起城市与生态功能上的连续性。

Enclosures allow for the cultivation of fruit trees, redcurrant bushes and hazel trees, whose fruit can be freely gathered by walkers. Elsewhere, meadows planted with standard fruit trees follow on from a sector of broom heath, then give way to a young woodland plantation.

Maturing into a natural space, the park evolves according to adapted management. It forms part of a deliberate approach to protecting and enhancing biodiversity through economical management. Only the edges of the paths are mowed regularly; the maintenance of the meadows is adapted to the flowering seasons and its use by the public. The hedges and drainage ditches are designed as ecological corridors, planted with hardy local species.

A similar dynamic has been developed in the planning of the eco-neighbourhood beside it: the landscape encroaches on the private lots. It allows for the organisation of different uses, and to ensure urban and ecological continuity between public and private spaces.

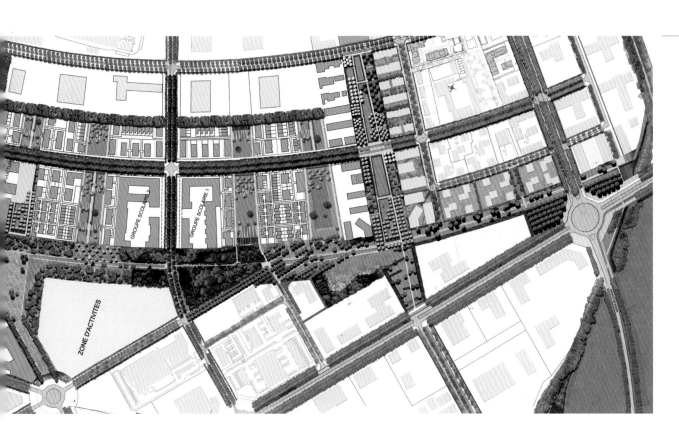

左页图：一个公园式住宅地块的等角透视图。
上图：街区配置图，以公园作为组织整个生态街区的结构性元素。

Opposite page: Axonometric diagram of a park block.
Above: Ground plan of the park structuring the eco-neighbourhood.

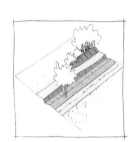 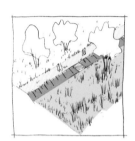  

上图：布里高原和马恩河畔坡地的景观，它们成为城市公园的组成元素。
右页图：公园中的大草原区，2011年7月。

Above: Landscapes of the Brie plateau and the Marne hillsides, the vocabulary of the urban park.
Opposite page: The large meadow in July 2011.

左页图：景观语汇与材料。
上图：果园区的植物混合，实践多样化管理的第一步。

Opposite page: Vocabulary and materials.
Above: The orchard – first act in the implementation of differentiated management.

上图：公园中的大草原区，2011年7月，草地收割策略的灵活运用。
右页图：果园区的一条小径。雨水的收集成为组织公园空间的重要考量之一。

Above: The large meadow in July 2011, showing its haymaking strategy.
Opposite page: A path through the orchard. Water collection underlies the park's organisation.

法国 塞尚/自2008起
Haie Cerlin eco-neighbourhood
赛尔兰树篱生态街区

方案的所在地位于塞尚，是南锡城乡区域最后的建筑储备用地之一，其开发目的在于提高城市密度和强化景观质量。

400户住宅围绕着绿化的院落布置，这些院落是对当地田园农舍院落形式在景观和环境上的一种重新诠释。住宅建筑坐北朝南以便节约能源，它们同时被安置在斜坡的垂直方向以利于地面雨水的收集和重新渗透。这些院落被设计成为居民的共享空间，具有社交性和实用性，并且成为城市中新"生态实践"的发展场所。

尽管该项目的建筑、城市和景观设计指导原则已经表明了在保护生态环境层面的远大目标，但是为了确保整体景观结构的持续性，土地在交付建造之前，整治管理单位已经预先在位于公共空间和私有地块之间的中介空间进行了绿化。

Situated in Seichamps, the sites for this project form one of the last reserves of unbuilt land in the greater Nancy area, and have been approached in terms of densification and the intensity of the landscape.

The 400 homes – facing south to maximise economy of energy and placed perpendicular to the slope to collect stormwater on the surface and allow it to percolate back into the ground – are organised around greens. These greens, places for neighbours to meet and use, are designed as shared spaces where the new "eco-practices" of the town will develop. Their landscape architecture is an environmental reinterpretation of the traditional farm layout in the villages of Lorraine.

Statements of Works define the environmental ambitions of the project in terms of architecture, town planning and landscape architecture. However, the planting of the boundaries between public and private space is already done when the plots are handed over to the construction firms in order to guarantee the future of the landscape structure.

左页图：生态街区的位置与范围。
上图：项目与城市的融合，整治方案的网络系统。建立城市、环境与景观的连续性。

Opposite page: The location and boundaries of the eco-neighbourhood.
Above: Urban insertion and the planning grid, a way of considering urban, environmental and landscape continuity.

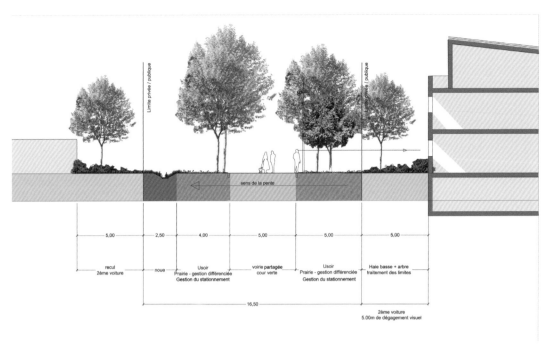

上图：一个绿化院落的剖面与透视效果图，当地传统农村空间的现代诠释。
右页图：私人空间围篱的植物种植原则和街区整体配置图。植物的种植方式在城市规划和设计中扮演着重要的角色。

Above: Cross-section and artist's impression of a green, showing the contemporary interpretation of the "usoir" (front yards open to the road in Lorraine villages).
Opposite page: Suggested planting for the hedges on the private plots. General development plan, showing how the town planning is based on the detail of planting.

Principe de plantation des haies

Principe / distance plantation

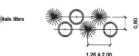

Haie libre

Haie libre de composition symétriques semi persistantes

Haie libre de composition symétriques semi persistantes

Haie libre de composition symétriques semi persistantes

Haie libre de composition asymétrique: grands arbustes au fond et petits arbustes devant

Haie taillée

Haie taillée entièrement caduque

Haie taillée 50/50 persistant/caduc

Haie taillée 50/50 persistant/caduc

Haie taillée 2/5 caduque 3/5 persistante

Haie taillée 1/3 caduque 2/3 persistante

Principe de plantation d'une haie double, multistrate

La multiplicité des troncs en cépée donne aux espaces plantés une forme de bosquet plus naturelle que celle des arbres tiges

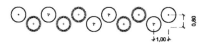

93

法国 俄德库尔与农萨尔 / 自2012起

Madine lake
玛蒂那湖

面积1100公顷的玛蒂那湖是一片储存饮用水的人工蓄水池，也是洛林地区的一个水上娱乐基地和观光景点，虽然具有一定的吸引力，但是在法国市场竞争中定位并不理想。整个区域拥有森林、湿地、草地、堤坝和沙滩，形成了独特的自然资源。

该景点的经济振兴必须通过非季节性的旅游业和一种更具有持续性的经济活动来带动。方案提出建立新的娱乐设施、扩建港口、提供多样化娱乐的住宿空间，同时加强景观特色。

这个项目与生态环境紧密结合，并开发了与大自然相关的游乐活动。

一个融合于景观之中且具有设计感的800米带状甲板，即所谓的"漫步建筑"，整合了所有的服务设施，并形成了此景点的入口意象和凝聚力；它同时也整合了各个散步道和景观视线，区隔了城市和自然，并彻底地更新了景点的形象。

这个方案在基地的尺度上对土地进行组织与塑造，犹如一项大地建筑的作品。

Madine lake, a 1,100-hectare artificial reservoir in Lorraine providing drinking water, is also a watersports and tourist park. Though attractive, it is badly positioned in relation to its competitors. In addition, its land, composed of forest, wetlands and meadows, dikes and beaches, forms an exceptional piece of natural heritage.

The economy of the site has to be reinvigorated with nature tourism that is less seasonal, and more consistent economic activity. The guide plan drawn up allows for the addition of new leisure facilities, a larger marina, accommodation and diversified leisure activities, while bringing out the specific qualities of the landscape.

The project is embedded in the environment and structures the natural leisure activities.

A "walkway building" – an 800-metre-long deck that forms a design line in the landscape – brings together the facilities, forms the gateway and the central feature of the site, is a launching point for walks, frames the views, shows where the urban ends and nature begins, and gives a radical new image to the site.

The architectural project for the site becomes a land-use project for the whole park.

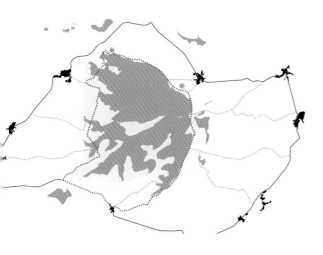
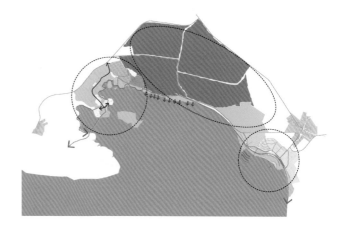

左页图：基地现况，方案构思的出发点。
上图：方案是基地与功能计划的融合结果。

Opposite page: The site today as the motivation for the project.
Above: The project as a fusion of a site and its future uses.

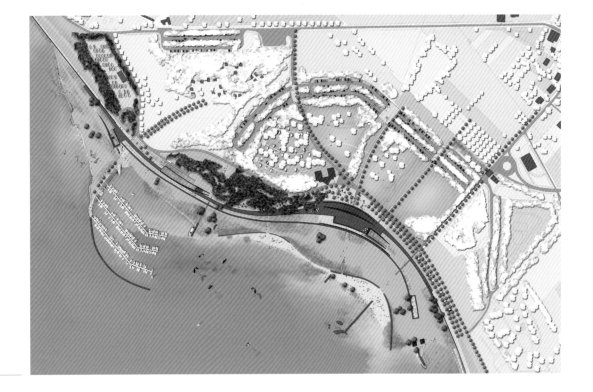

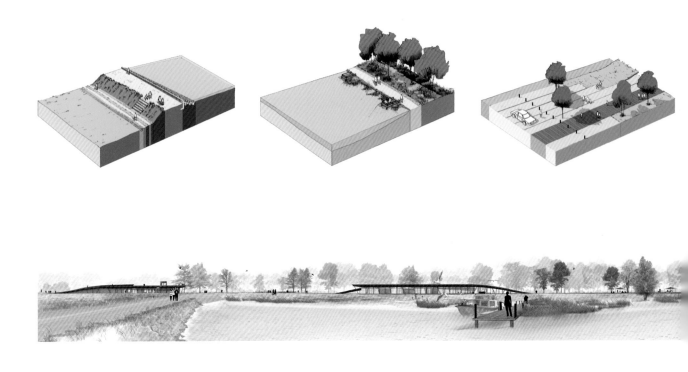

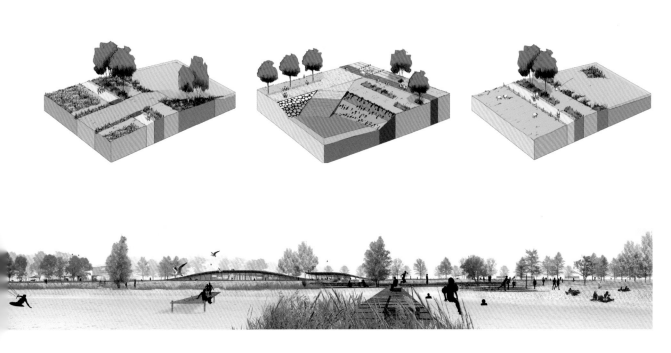

一个拥抱湖面景观并融合了建筑物的散步空间。　　　　　　　　　　A walk that is open to the lakeside landscape and integrates the buildings.

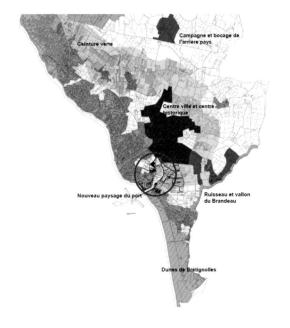
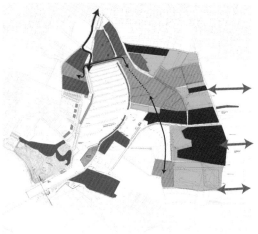

法国 滨海不列提纽勒 / 自2009起

yachting marina
不 列 提 纽 勒 游 艇 码 头

不列提纽勒位于旺代地区的海岸线上，迫于大量游客的压力，小城试图减缓别墅住宅区的大量开发，通过兴建游艇码头来发展新的地区性经济。本项目开辟了一条"绿带"来限制城市扩张，将某些建设用地转变成树林，并且着手规划一个港口。

这个新建设的10公顷水面区域可容纳1300艘游艇停靠。一个位于基地内的采石场被重新开发，以提供建造防冲堤所需的石材，随后再以挖掘水塘所产生的土方来回填整平。

方案在环境保护的层面上独具匠心，试图最大程度地降低它对周围自然空间的影响（例如：开挖最小面积的水道、砌筑防冲石堤、建造具有透水性地面的停车场，重新利用采石场等措施）。

方案计划在未来码头的周围建立一个与城市相连接的建筑立面。水塘四周的空间则反映出基地的景观特色，并将其延伸：为沙丘重新注入自然活力、对树篱和湿地系统进行保护，并且种植了几公顷幼林。

Situated on the Vendée coast and popular with tourists, Brétignolles-sur-mer is betting on halting suburban sprawl and creating a yachting marina to develop a new local economy. It has put in place a "green belt" to limit the spread of urbanisation; building plots will be transformed into woods and a port project is being developed.

The 10-hectare body of water created will provide mooring for 1,300 yachts. A quarry was reopened on the site to provide the riprap necessary to build the structures, before being backfilled by the earthworks generated by the basin.

The port project is ambitious from an environmental point of view, and seeks to minimise its impact on the natural spaces (minimal channel width, the use of riprap, parking on permeable ground, reusing the quarry, etc.).

A waterfront that prolongs the town is planned around the future port. The spaces around the basin are inspired by the landscapes of the site and perpetuate them: the dunes are returned to nature, the small fields and hedges and the wet meadows are protected and several hectares of young woodland have been planted.

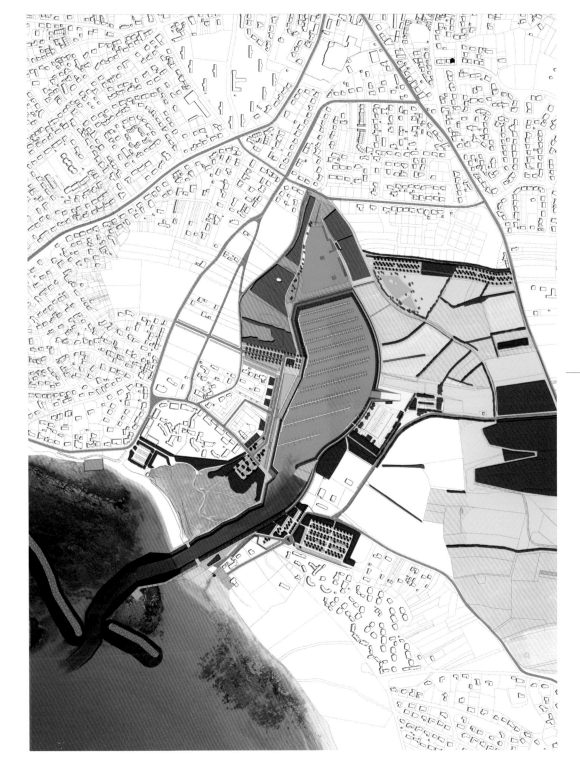

左页图：种植林木的绿化带限制了城市扩张，以便发展新的港口经济。
上图：港口设施配置图。

Opposite page: The wooded green belt that halts urbanisation in order to create a new port economy.
Above: Ground plan of the port structure.

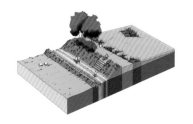
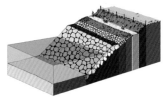

左页图：地面整治填挖的管理。
上图：港口方案成为重新建立自然生态环境与改善景观的良机。

Opposite page: Management of the dredged waste and the backfill.
Above: The marina offers the opportunity to reconstruct natural environments and to rehabilitate the landscape.

102

左页图：城市码头与水道。
上图：船坞的土岸与海水浴场。

Opposite page: The town quay and the channel.
Above: The marina from the land side and the bathing beach.

法国 立梅 / 自2005起
Limay port
立 梅 港 口

立梅港口位于塞纳河畔，是一个拥有商业和重工业活动的多功能区域。接续着帕特里克·科隆比尔（2000－2004年该港口的顾问建筑师）的研究工作，我们在2005年完成了一项应巴黎港口管理局委托、针对港口整治和可持续性发展的整治纲要，成为港口发展的指导原则，其中包括了有关经济发展、生态环境保护与发展、土地整治与使用等方面的策略。

此整治纲要的重要挑战在于：在推动河流运输发展的同时，也要证明港口的活动可以以正面的方式融入所在的地域当中，可以生成正面的景观和高质量的环境。

这份整治纲要同时是一个策略计划、一个为期多年的行动计划、一个针对实施项目的管理方法以及一个景观和城市规划的基础依据。

The river port of Limay, on the Seine, is a multifunctional zone hosting commercial activities and heavy industry. In 2005, following on from the work of Patrick Colombier (the port's consultant architect between 2000 and 2004), a land-use and sustainable development plan was drawn up by the Ports of Paris. It was to become the port's guide plan in terms of strategies for economic development, enhancement and environmental protection, land-use planning and the occupancy of the area.

The challenge of this development strategy is to drive the development of river traffic by showing that the port activity can fit into its territory in a positive way, generating a beneficial landscape and a quality environment.

The development plan is both a strategic plan, a plan of action over several years, a means of managing the arrival of new businesses building their facilities on the site, and a town-planning and landscape document.

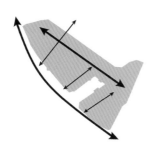
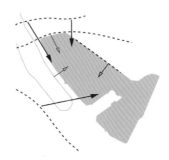

左页图：立梅港口现状，必须加以保留与改善，并解决塞纳河在大巴黎发展计划中所面临的城市与自然如何共处的问题。
上图：立梅港口空间整治与可持续发展的纲要，一个同时涉及城市、景观与环境的指导原则。

Opposite page: The views of the port, both distant and in proximity had to be maintained and strengthened – the Seine in Greater Paris exemplifies the confrontation of the city and nature.
Above: The Land-Use and Sustainable Development Plan for the Port of Limay, a guide plan for the city, the landscape and the environment.

为了付诸实施，此纲要提出了多项作为规划依循的资料，包括土地发展计划、公共空间指导计划、自然绿色空间管理计划、种植与自然化计划，以及城市、建筑、环境和景观的规划指导。此外，"绿色计划"制定了年复一年的种植活动，以及不同空间所需要的不同频率的维护计划；农业类型的管理方式受到重视和倡导，以保持最低的维护成本。

按照土地征用，众多大块土地被纳入管理计划。这些地块受到播种、种植和维护，逐渐显现出自身的特点，并成为港口的一部分。港口景观因而逐渐形成，交错着工业设施以及以树篱作为边界的广阔草地。

To these ends, several documents were prepared: a plan for land development, a guide plan for the public spaces, a management plan for the green and natural spaces, a plan for planting and restoring denatured sites, and a Statement of Works covering urban, architectural, environmental and landscape specifications. "Green plans" programme the planting campaigns year on year and class in order of priority and frequency the maintenance of the different spaces. An agriculture style of management is favoured in order to minimise the costs of upkeep.

As land is acquired, broad parcels are integrated into the management plan. They are then sown, planted, maintained and become identifiable as belonging to the port. Progressively, the landscape of the port is constructed: industrial infrastructure is cheek by jowl with wide hay meadows, framed by hedgerows.

特别值得一提的是，建筑规划指导原则中建立了一个优先应用于港口建筑和构筑物的色彩规范，通过一系列鲜明亮丽的颜色，将港口具有特色的建筑元素凸显出来（储藏塔、货柜、烟囱等）。

随着时间流逝，随着新的建设和新的行动，在长达9年的景观和城市规划任务的进行下，一个合乎构思计划的工业景观渐渐成型，一个协调合宜的景观被建构起来，生态环境也得到了改善。这些策略的持久性让立梅港口成为可持续发展的一个典范。

The architectural specifications include, among other things, a colour chart to guide the choice of colours on the buildings and smaller structures of the port. A range of bright and lively colours puts into relief certain architectural elements characteristic of port activity (silos, containers, chimneys, etc.).

Under the aegis of a nine-year town-planning and landscape mission, new buildings and operations have arrived, a controlled industrial landscape has taken shape, the structure has gained coherence and the environment blossomed. The permanent impact of these strategies makes the port of Limay a model for sustainable development.

索恩河畔沙隆的布莱-得旺公园。   Prés-Devant Park in Chalon-sur-Saône.

新景观经济
a new economy of landscape

掠夺性的经济所导致的自然或人为灾难（！？）近年来变得愈加频繁，这些现象是景观设计的重要课题。

河水泛滥其来有自，城市中过多的不透水硬质地面造成了雨水无法排除，而受到城市压力日渐缩小的河床已经无法再容纳吸收过多的雨水。雨水管理的替代方式成为许多城市设计的共同课题，并为城市空间整治带来新的形式语汇。

相对于经常使用农药产品的乡村来说，城市中的生物种类更容易多样化。绿化空间多样化管理的实施减低了这些空间在维护上的压力，尽管人们经常对这些提倡"杂草"的策略感到不甚理解。

如今的河流演变成毫无生机的渠道，规划设计的目的便在于重新赋予河床蜿蜒曲折的形式，运用土木工程中的软性技术来修复河岸，以重现河流物种丰富多样的生态面貌。此外，我们也以相同的思路来设计雨水排水工程。

随着城市的发展，无法分解的废弃物也伴随而来，花园的植物层可以掩盖这些不可回收的垃圾，为处理这类污染提供了适当的方法。

游客过量是自然风景区域衰败的主要原因，游客流量的管理和重新自然化是景观设计师所采用的新实用科学。港口项目对发展运输与娱乐经济非常重要，然而，若非通过景观方案来进行构思，这些目的终究无法达成。

经济通常被视为是带有风险的活动，然而发展项目所带来的财富，应在前期阶段就运用一部分在平衡环境与社会因素方面，以更好的条件让经济发展融入景观当中。

Natural or human (!?) risks, induced by a predatory economy, are becoming more and more prevalent as the themes of landscape projects.

Rivers flood because their beds have been reduced by urban pressure and no longer absorb the excess water from the all-too-common land sealing. Alternative water management has become commonplace in urban projects, and forms the basis for a new development vocabulary.

Urban biodiversity has more chance of being regenerated than that of our countryside, which is pulverised by phytosanitary products. Differentiated management of the green spaces, decreasing the pressure of upkeep, is implemented together with an educational message about "weeds" in order to combat the frequently encountered lack of understanding on the part of the inhabitants.

Rivers that have become sterile conduits are the subjects for a reconquest of biodiversity space by giving riverbeds back their meanders and by restoring the banks through the soft technologies of civil engineering. We also design stormwater drainage works as river projects.

As the city moves forward, we discover discharges that are impossible to reabsorb. Gardens then become opportune for confining this irreversible pollution under a plant layer.

Over-visited natural sites are their own worst enemy. Tourist flow management and restoring these sites to nature are new applied sciences used by the landscape architect. Port projects, so necessary to the transport and leisure economy, need to be landscape projects to have a chance of reaching fruition.

The economy is perceived as a risk, but the benefits brought about by the economic development project can in part be directed upstream, to balancing out environmental and social factors and laying a better ground for its integration and acceptance in the landscape.

法国 滨海卡约 / 自2006起

Hourdel point
乌 戴 尔 海 角

乌戴尔海角是索姆海湾南部的突出部分，它独特的景观在每年的旅游旺季吸引了大批游客。这个高强度的观光业逐渐损坏了海角的自然环境，一种介于沙丘和淤泥之间的微妙平衡。

将乌戴尔海角纳入一个"大基地计划"的构想需要三个因素之间的协同效应来促成，包括一个具有吸引力的地方经济、对高品质海滨自然环境的保护，以及土地和观光频率之间关系的协调。

想要获得改善和保护的自然空间必须与游客隔离。通过对游客的疏导，使他们与环境之间保持适当距离，有利于形成人们对所造访的空间的尊重。

方案提出了在一个大尺度范围内对交通流量和停车进行合理的管理，同时对乌戴尔村庄和它周围的自然空间进行改善。交通策略的研究是根据旺季和淡季的游客流量来进行的，同时伴随着一个将小村庄本身需求纳入考量的停车规划。

一些小路将游客从停车空间带往广阔的景观，使他们循序渐进地发觉基地的景致、产生无限的遐想，并使他们以一种感性的方式感受这片土地。

Hourdel Point closes the southern part of the bay of the Somme. Its exceptional landscapes attract large numbers of visitors each year at high season. This intensive tourism is gradually deteriorating the natural environments of the point, which rely on a subtle balance between the dunes and the mud flats.

The decision to become designated as a "major site" necessitates finding a synergy between an attractive local economy, preserving of the value of the natural coastal environments and reconciling the district with the numbers of visitors it attracts. The natural site must be removed from the public in order to restore and protect it. This distance encourages respect for the area frequented by visitors and a channelling of tourist activities.

A well thought-out, large-scale plan for the management of traffic flow and parking is proposed, together with an upgrading of the hamlet of Hourdel and the natural space that surrounds it. The traffic strategy was planned taking into account the periods of high and low tourist attendance. With it comes a parking strategy that respects the use of the hamlet.

Footpaths link the parking areas to the larger landscape. They encourage a slow and progressive discovery of the site, invite visitors to dream and to approach the area with sensitivity.

乌戴尔海角的整治方案必须思考如何在旅游业带来的压力和自然空间保护之间取得平衡。

Hourdel Point: How to reconcile tourist pressure and the preservation of natural spaces.

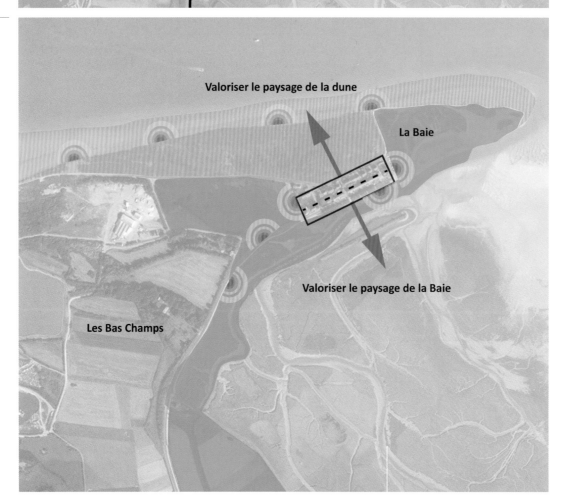

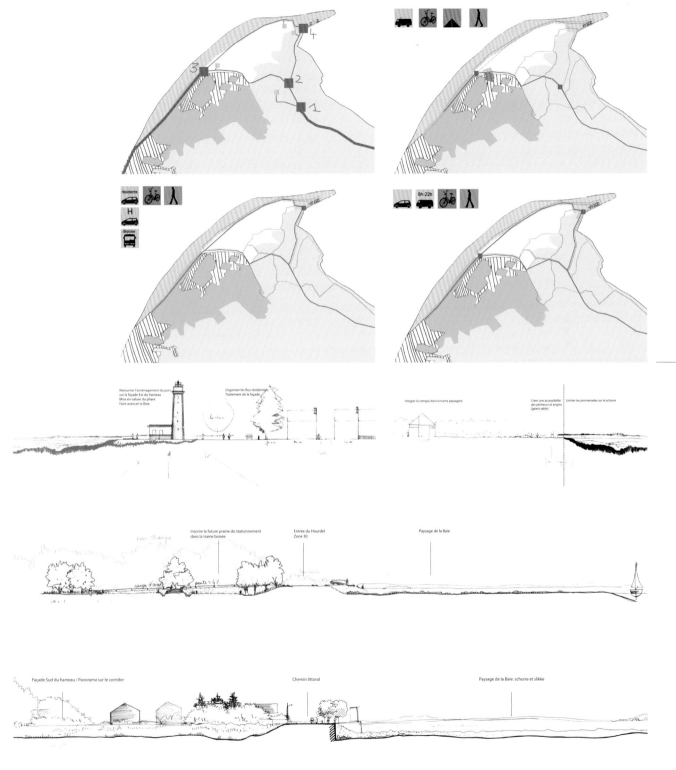

左页图：旅游管理策略规划和自然空间保护与改善计划。
上图：整治方案里面的各种空间序列。

Opposite page: Strategic plans for the management of the flow of tourists and making the most of the natural spaces.
Above: Development sequences.

法国 贡夫勒维尔洛谢 / 自2004起
Saint-Laurent River
圣洛朗河

小村庄古尔内坐落在一个下陷的谷地，经常规律性地被埋入地下管道的圣洛朗河水、废弃的工业排水渠以及由不可渗透的土质所聚集的雨水所淹没。2004年发起的竞赛方案建议在河谷底部重新建立河床来解决水流的问题。

这个需要对空间进行重新整治的水利工程同时也是一个城市和环境的改造工程，必须在河水流经之地重新组织公共空间和交通、合并土地、建造公共设施，并且拆除或者迁移住宅建筑。

Situated in a valley with steep escarpments, the hamlet of Gournay is regularly flooded by the Saint-Laurent River culvert, by its abandoned industrial reach, and by the discharge from high plateaux that have lost their permeability. A competition launched in 2004 called for bids to resolve the flooding problems by creating a new river bed at the lowest point of the valley.

This hydraulic project, which calls for a restructuring of the area, is thus both an urban and an environmental project. Along the river's route public spaces and traffic circulation has to be reorganised, land lots regrouped, public facilities rebuilt and houses destroyed or relocated.

古尔内村庄和圣洛朗河。解决涨潮洪泛的问题成为这个重整城市空间项目的原始动机。

The hamlet of Gournay, the course of the Saint-Laurent and its reach. Combatting flooding becomes an incentive for urban restructuring.

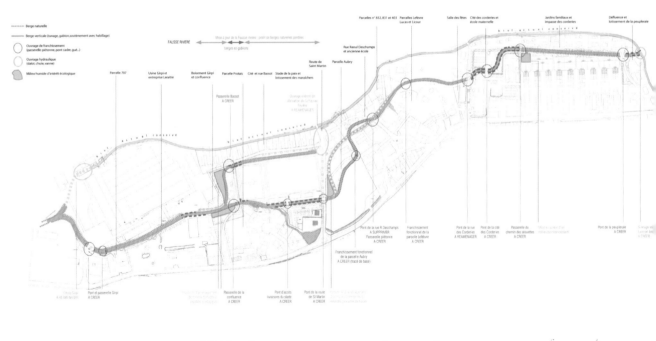

最初，方案是朝零风险的方向展开研究，但考虑到水利项目无法保证完全的防护，之后就朝向合理风险管理的方向发展。由于不能完全解决洪水的问题，方案的研究重点于是放在减轻洪水带来的后果。如今，地方政府买下那些可能会被淹没的住宅，将其进行改造、底层架空。圣洛朗河则被整治成更加自然的形态，能够快速地排水引流，其河道不再是具有防洪作用的工程设施，而是一个可以调节自然风险的自然空间。

通过众多协议讨论以及地方政府的大力投入，当地居民紧密地参与了这项环境的改革整治和城市的巨大变化。土地所呈现的现实面以及随着自然现象而产生的社会演变渐渐地成为方案发展的重要考量。

Initially aimed at finding a zero risk solution, the hydraulic programme has evolved towards a reasonable management of the risks after studies found that total protection was not be possible. The project no longer combats flooding but aims to minimise its consequences. The community now buys back the floodable houses to rebuild them on stilts, and the river has been landscaped with a more natural profile that can evacuate the water rapidly. The water course is no longer an anti-flood construction, but a natural space that "manages" a natural risk.

Through a strong programme of consultation, with major investment by the local authorities, the population is closely involved with this environmental revolution and urban change. Slowly, the project has fitted into the reality of the area and society's evolution in relation to natural phenomena.

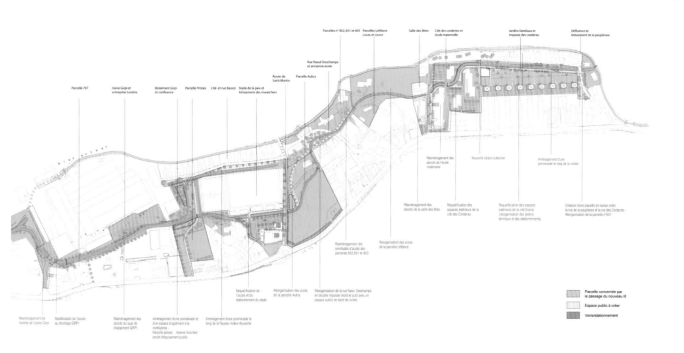

左页图：五十年一遇洪泛区域分布，以及不同河岸类型的建议。
上图：跟随新河道整治所产生的各式公共空间。

Opposite page: The 50-year floodmark, and the different bank typologies proposed.
Above: The different projects for public spaces brought about by the new river.

法国 索恩河畔沙隆 / 自2006起
Prés-Devant suburb
布莱－得旺市郊

索恩河畔沙隆新医院的设立和城市环线的完成迫使人们重新对其市郊布莱-得旺的发展进行一个追溯既往的整体性思考，包括其公共空间的重新定位、交通流量的规划、医学设施的发展、被随意丢弃之垃圾废物的回收、对蔓延性植物的防范、对塔利河涨潮的控制，以及对项目有限预算的管理。

从前的省级公路变成了公园的林荫道，废物的集中地产生了新的地貌，塔利河的湿地形成了花园，而公园则成为可吸收洪泛的地区。各种限制和未被解决的问题成为项目设计的新动机，创造了一些平凡无奇却起到了正面作用的新景观。

The building of Chalon-sur-Saône's new hospital, together with the completion of the bypass, imposed a need for retroactive thinking about the whole suburb of Prés-Devant. On the table were: redefining the public spaces, traffic flow, the development of a medical cluster, reducing uncontrolled landfill, combatting invasive plants and flood management for the River Thalie, all within a very limited project budget.

The old departmental road becomes the main walk through the park; the containment of the landfill creates a new topography; the Thalie wetlands become gardens; and the public park is floodable. Taking the constraints and residual problems as the basis for a project generates a new landscape, normalised and positive.

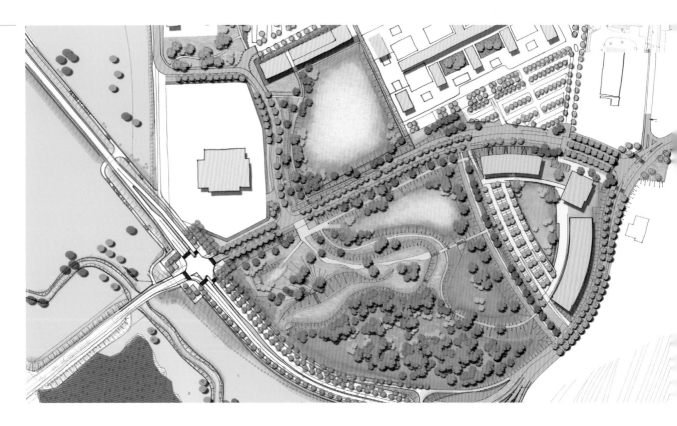

布莱-得旺公园和谐地融入了塔利河谷的农业景观中，能够容纳河流的涨潮泛滥，并赋予了新医院街区一个景观上的特色。

Prés-Devant park fits into the agricultural landscape of the Thalie valley, allows for the expansion of the river under flood and gives the new hospital neighbourhood its identity.

方案的两个施工阶段。
上图：进行去污染处理时期的状态。
下图：原先的省级公路转化成公园的林荫道，废物集中地变成了池塘。

Two stages of the site under construction.
Top: The Prés-Devant sector during its clean-up.
Above: The road becomes a walk through the park and the landfill a lake.

上图：作为背景的湿地景观。
右页图：公园景观，其中包含一个冬季涨潮景象。

Above: The wetland landscape in the background.
Opposite page: The park, including during a winter flood.

123

上图：基地在2008年的状态以及经过整治后的现状，背景为沙隆医院（建筑师 Agence Brunet Saunier）。
右页图：公园景致，一个城市中的自然空间。

Above: Before work began in 2008, and after completion. In the background is Chalon hospital (architects: Agence Brunet Saunier).
Opposite page: The park, a natural urban space.

125

法国 赛里 / 2001

Gassets brook
格赛小溪

格赛小溪是布利亚尔高原溪流的一部分，也是赛里镇的空间结构要素。改造项目最初计划在公共空间地下设置水流管道，将小溪整治成地下化的卫生工事。

方案的最大挑战在于说服作为业主的公家整治机构让溪流维持露天状态，并建立一些与其相关的公共空间。因而方案建议以其他替代方式来进行雨水处理，同时将小溪沿岸改造成在洪泛时期可被淹没的散步平台。

小溪和各公共空间之间巨大的高差展现了多种河岸处理手法：斜草坡、柴笼或者石笼。从"自然状态"的溪流到注入防洪水池之间的整段流线被细心处理成不同的景观片段。

小溪以错落有致的观景露台和向水面倾斜的船坞来与城市产生对话，一些小径和台阶让人们可以穿过溪流或者在岸边停留休憩。

沿河的树林和湿地草场发展出独特的生物多样性，形成了"城市中的自然"。然而，最近一段时间，对植物的过度维护使生态环境逐渐变得贫瘠，于是方案也建立起一个适当的管理计划以伴随着这个城市自然空间今后的发展。

Gassets brook, a stream running down from the Briard plateau, structures the town of Serris. The development programme envisaged lining the stream and channeling it underground to transform it into a wastewater facility beneath the public space.

The main challenge of the project was to convince the local authority to keep the stream above ground and to enhance it by the creation of connected public spaces. The project therefore suggests an alternative stormwater management system and transforms the banks into a floodable walk.

The significant gradient between the stream and the public invites a varied vocabulary of talus, willow fascines and gabions. Several sequences create a journey from the "natural" stream to its outfall into a large storm basin. The stream opens onto the town through a series of lookout points and holds sloping towards the water. Paths and flights of steps allow the public to go down to the brook and sit on its banks.

Riverside vegetation and wet meadows encourage a specific biodiversity, producing an effect of "nature in the town". Recently, however, over-intensive management of the vegetation has undercut the richness of the environment. This project proves to what point the question of management is vital. By drawing up a management plan it will be possible to follow and encourage its evolution over time.

溪边地面以防止渗透的都市化形式处理后,产生了新的景观与环境。

The project generates a landscape and an environment born of the waterproofing of urban ground.

左页图：防洪蓄水池旁的观景台。
上图：大阶梯与整治细部。项目从原本的卫生工事转变为一个城市公共空间的整治方案。

Opposite page: The lookout point over the storm basin.
Above: Steps down to the brook and details of the restructuring work. A wastewater project turned urban project.

方案索引

projects index

以下资料中的造价为不含税价格 The following construction costs are calculated excluding VAT.

非尔特尔与卡勒瓦尔街 pp.014-019
FELTRE AND CALVAIRE STREETS
法国 南特 Nantes, France – 2007
合作者 / With : Sogreah BET VRD
业主 / For : Nantes Métropole
12 500 m² – 200 €/m²

欧洲大道 pp.020-025
EUROPE AVENUE
法国 大莫特 La Grande Motte, France – since 2011
合作者 / With: Safège BET VRD, Transitec Circulation, ON Concepteur Lumière, F. Azambourg Design
业主 / For : City of La Grande Motte
28 000 m² – 250 €/m²

阿里亚娜广场 pp.026-029
ARIANE SQUARE
法国 谢西 Chessy, France – 2002
合作者 / With : Setec TPI BET VRD
业主 / For : EPA Marne
20 000 m² – 150 €/m²

洛克瑞斯特市中心 pp.030-035
LOCHRIST TOWN CENTRE
法国 安赞扎克-洛克瑞斯特 Inzinzac-Lochrist, France – 2007
合作者 / With : Sogreah BET VRD
业主 / For : City of Inzinzac-Lochrist
40 000 m² – 80 €/m²

奥斯特利兹堤岸 pp.036-039
AUSTERLITZ QUAY
法国 巴黎 Paris, France – since 2007
合作者 / With : BATT BET VRD, ON Concepteur Lumière
业主 / For : Ports of Paris
20 000 m² – 230 €/m²

纪龙德河堤岸 pp.040-045
GIRONDE QUAYS
法国 波亚克 Pauillac, France – since 2010
合作者 / With: Safège BET VRD, Artkas Scénographie et Signalétique, Scène Publique Concepteur Lumière
业主 / For : City of Pauillac
95 000 m² – 100 €/m²

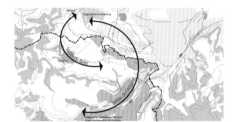

贝雷尔城门 pp.046-051
BELER GATEWAYS
法国 惠桑泽平原 Plaine de Russange - CCPHVA, France – 2009
合作者 / With : Y.L Aménagement, Ingérop BET VRD
业主 / For : Community of Communes of Pays Haut Val d'Alzette
180 ha

圣夏尔矿井 pp.052-055
SAINT-CHARLES PIT
法国 小罗塞尔 Petite Rosselle, France – 2010
合作者 / With : Séfiba BET VRD
业主 / For : EPF Lorraine and City of Petite Rosselle
27 ha

万德乌市中心 pp.058-063
VANDŒUVRE TOWN CENTRE
法国 万德乌-雷-南锡
Vandœuvre-lès-Nancy, France – since 1998
合作者 / With : Séfiba BET VRD
业主 / For : Urban Community of Grand Nancy, City of Vandœuvre-lès-Nancy, Batigère
13,5 ha – 11 M€

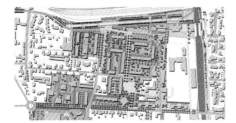

维特里郊区   pp.064-069
VITRY SUBURBS
法国 维特里-勒-弗朗索瓦 Vitry-le-François, France – since 2004
合作者 / With : Séfiba BET VRD
业主 / For : Community of communes of Vitry-le-François, Vitry Habitat
22 ha – 81 €/m²

梅兹十字街区   pp.070-075
CROIX DE METZ NEIGHBOURHOOD
法国 图勒 Toul, France – since 2006
合作者 / With : Séfiba BET VRD
业主 / For : City of Toul, Toul Habitat
20 ha – 56 €/m²

塞纳河畔丘陵地   pp.080-081
SEINE HILLSIDES
法国 安德莱西 Andrésy, France – 2009
合作者 / With : Y.L Aménagement, AVR BET VRD
业主 / For : EPAMSA, Commune of Andrésy
110 ha

艾夫兰山公园   pp.082-089
MOUNT ÉVRIN PARK
法国 蒙泰夫兰 Montévrain, France – since 2005
合作者 / With : Prolog Hydrologie, ATPI Infra BET VRD, Zoom Écologue, Pixelum Lumière, Gras et Miroux (eco-neighbourhood)
业主 / For : EPA Marne, City of Montévrain
20 ha – 40 €/m²

赛尔兰篱生态街区   pp.090-093
HAIE CERLIN ECO-NEIGHBOURHOOD
法国 塞尚 Seichamps, France – since 2008
合作者 / With : Sefiba BET VRD, Bepg BET Hydraulique
业主 / For : Solorem
21 ha – 40 €/m²

玛蒂那湖   pp.094-097
MADINE LAKE
法国 玛蒂那湖 Madine, France – since 2012
合作者 / With : 2AD Architecture, Safège BET VRD
业主 / For : Mixed Syndicate for the Development of Madine Lake
20 ha – 9 M€ with the buildings

不列提纽勒游艇码头   pp.098-103
YACHTING MARINA
法国 滨海不列提纽勒 Brétignolles-sur-mer, France – since 2009
合作者 / With : BRL mandataire, Arcadis Ingénierie, Biotope
业主 / For : City of Brétignolles-sur-Mer
90 ha

立梅港口   pp.104-111
LIMAY PORT
法国 立梅 Limay, France – since 2005
合作者 / With : Ernst & Young Consultant
业主 / For : Ports of Paris
125 ha

乌戴尔海角   pp.112-115
HOURDEL POINT
法国 索姆海湾 Bay of the Somme, France – since 2006
合作者 / With : Geodice Déplacements, Zoom Écologue, Infraservices BET VRD
业主 / For : Mixed Syndicate of Bay of the Somme Grand Littoral Picard – 40 ha

圣洛朗河   pp.116-119
SAINT-LAURENT RIVER
法国 贡夫维尔洛谢 Gonfreville-l'Orcher, France – since 2004
合作者 / With : Sogreah, BET Hydraulique
业主 / For : City of Gonfreville l'Orcher - Agglomeration Community of Havre
42 000 m² – 200 €/m²

布莱-得旺市郊   pp.120-125
PRÉS-DEVANT SUBURB
法国 索恩河畔沙隆 Chalon-sur-Saône, France – since 2006
合作者 / With : PMM BET VRD et infrastructure, Cabinet Reilé BET Environnement
业主 / For : SEM de Bourgogne
12 ha – 40 €/m²

格赛小溪   pp.126-129
GASSETS BROOK
法国 赛里 Serris, France – 2001
合作者 / With : Tugec Ingénierie BET VRD
业主 / For : EPA Marne
6 600 m² – 130 €/m²

事务所简介

biography-agency

让马克·高里耶（1958年出生）是城市规划师、景观设计师与建筑师，在1996年成立了宇比库斯事务所，发展出"大地建筑"的想法，成为建筑师-景观设计师实践景观专业的一种方式。

在1985年和1995年之间，他受到亚历山大·谢梅道夫的教学启发和其景观办公室所提供的实务经验，使得他获得了"法国景观联合会认证之景观设计师"的头衔。1998年，国立布卢瓦自然与景观学校（ENSNP）聘请他为教授，先后担任了四年级"大型景观"方案教学以及三年级城市规划工作室的联合负责人。2010年，他带着城市规划和景观方面的资深经验，成为国家城市更新机构（ANRU）专家组的一员。

如今，拥有着16年历史和20名员工的宇比库斯事务所仍然持续不断地发展着，以便将其经验转化为丰实的专业资本，并增长经验分享的能力，同时也通过专业实践，为建筑师-景观设计师在土地规划的领域中建立出更明确而具有影响力的角色。

Jean-Marc Gaulier (born in 1958) is a town-planner, landscape architect and chartered (DPLG) architect. He created the Urbicus agency in 1996, developing a practice that sees itself as an "architecture of the land", driven by landscape architects.

Between 1985 and 1995, at Alexandre Chemetoff's Bureau des Paysages (Office of Landscapes), he learnt by example and practiced the project, leading to his acceptance as a chartered landscape architect by the French Landscape Federation. In 1998 he began teaching at the National School of Higher Studies in Nature and Landscape Architecture in Blois (ENSNP), where he was successively co-director of the 4th year project course on "major landscapes", then of the 3rd year workshop on urban projects. In 2010, thanks to his experience as a landscape architect-town planner, he joined the National Agency for Urban Renewal (ANRU) centre of expertise.

Today, 16 years on and with 20 team members, Urbicus continues to structure itself to capitalise on its experience, and to increase its capacity to share it, and to claim, through practice, a more structuring role for landscape architects in land-use planning.

contributions 致 谢

以下人员为宇比库斯事务所目前的设计主力，参与各项方案的构思：

Tanja Aubourg, Claire Bellet, Rudy Blanc, Thomas Boisdet, Rostom Chikh, Olivier Courtelle, Amandine Doucet, Thimothé Dumas, Sandrine Feutry, Véronique Fourteau, Simon Gaulier, Sylvie Gourio, Évelyne Henriot, Noëlle Madrona, Bernadette Muchenberger, Solène Quilin, Chrystelle Rouge, Amina Zehouani.

以下合作伙伴对于本事务所方案(尤其本书所呈现的项目)付出了诸多心力：

Johanna Almgred, Yuli Atanassov, Grégoire Bassinet, Clément Bollinger, Élodie Bortoli, Marc de Verneuil, Federica Dell'Orto, Antoine du Plessis d'Argentré, Bénédicte Dufieux, Agathe Gresset, Guillaume Hugon, Soizic Larpent, Gaylord Le Goaziou, Marielle Lévy, Yohann Maillard, Luc Mallet, Maud Martzolf, Bénédicte Morel, Florent Morisseau, Grégory Ouint, Nicolas Renard, Jérôme Saint Chély, Olivia Samit, Laurence Sciascia, Lionel Sikora, Rémy Turquin, Anne-Sophie Verriest.

本书的筹划与编辑特别归功于：

Sandrine Feutry 和 Chrystelle Rouge

感谢所有人的贡献，在此展现了他们的才能、工作成果与热忱。

Currently working on Urbicus projects:

Tanja Aubourg, Claire Bellet, Rudy Blanc, Thomas Boisdet, Rostom Chikh, Olivier Courtelle, Amandine Doucet, Thimothé Dumas, Sandrine Feutry, Véronique Fourteau, Simon Gaulier, Sylvie Gourio, Évelyne Henriot, Noëlle Madrona, Bernadette Muchenberger, Solène Quilin, Chrystelle Rouge and Amina Zehouani.

Contributors to the projects presented in this book and other projects:

Johanna Almgred, Yuli Atanassov, Grégoire Bassinet, Clément Bollinger, Élodie Bortoli, Marc de Verneuil, Federica Dell'Orto, Antoine du Plessis d'Argentré, Bénédicte Dufieux, Agathe Gresset, Guillaume Hugon, Soizic Larpent, Gaylord Le Goaziou, Marielle Lévy, Yohann Maillard, Luc Mallet, Maud Martzolf, Bénédicte Morel, Florent Morisseau, Grégory Ouint, Nicolas Renard, Jérôme Saint Chély, Olivia Samit, Laurence Sciascia, Lionel Sikora, Rémy Turquin, Anne-Sophie Verriest.

This book was made possible by:

Sandrine Feutry and Chrystelle Rouge

Our thanks to everyone for their professionalism, their work and their motivation.

credits

版权说明

文字、图片与照片：© Urbicus

以下资料除外

A3 Production – p.44 上面四张照片
Agence Com'Air – p.121
镌印作品(私人收藏) – p.30 左上
旧明信片(私人收藏) – p.40 左
Jean-François Chapuis / Photothèque Smac – p.12, p.15 上
Éric Giretti – p.94
Ville de Gonfreville L'Orcher – p.117 上
Gras Miroux Architectes associés – p.82
Image in Air – p.101 上
François Marchand / Balloïde-photo – p.30 右, p.32-33 下
Éric Morency – p.27, p.126
Petite Rosselle 旧照片(私人收藏) – p.55 上
Sogreah – p.118 上
Terre d'images – p.37

版权属于本事务所（© Urbicus）的照片当中，一部分由以下人员拍摄：
Charles Delcourt 和 Michel Reuss

透视效果图的制作归功于以下人员：
Cube
Jean Joyon
Gaël Morin
Chloé Sanson

Texts, images and photographs: © Urbicus

except

A3 Production – p.44 four photos at the top
Agence Com'Air – p.121
Print (private collection) – p.30 top left
Old postcards (private collection) – p.40 left
Jean-François Chapuis / Photothèque Smac – p.12, p.15 top
Éric Giretti – p.94
Ville de Gonfreville L'Orcher – p.117 top
Gras Miroux Architectes associés – p.82
Image in Air - p.101 top
François Marchand / Balloïde-photo – p.30 right, p.32-33 bottom
Éric Morency – p.27, p.126
Old photographs of Petite Rosselle (private collections) – p.53 top
Sogreah – p.118 top
Terre d'images – p.37

Part of the photographic images credited to Urbicus was taken by:
Charles Delcourt and Michel Reuss

The perspective drawings are by:
Cube
Jean Joyon
Gaël Morin
Chloé Sanson

芒特拉若利的露天绿地剧场，2006年。

Open-air theatre in Mantes-la-Jolie, 2006.

图书在版编目（CIP）数据

景观与城市转变：宇比库斯事务所设计作品专辑 / 法国亦西文化编；陈庶译 . -- 沈阳：辽宁科学技术出版社，2012.9
ISBN 978-7-5381-7608-7

Ⅰ. ①景… Ⅱ. ①法… ②陈… Ⅲ. ①城市景观—景观设计—作品集—法国—现代 Ⅳ. ①TU856

中国版本图书馆CIP数据核字（2012）第170208号

出版发行：辽宁科学技术出版社
（地址：沈阳市和平区十一纬路29号　邮编：110003）
印　刷　者：利丰雅高印刷（深圳）有限公司
经　销　者：各地新华书店
幅面尺寸：210 mm×230 mm
印　　张：8.5
插　　页：4
字　　数：50千字
印　　数：1~2000
出版时间：2012年9月第1版
印刷时间：2012年9月第1次印刷
责任编辑：陈慈良
封面设计：维建·诺黑
版式设计：卡琳·德拉梅宗
责任校对：周　文

书　　号：ISBN 978-7-5381-7608-7
定　　价：88.00元

联系电话：024-23284360
邮购热线：024-23284502
E-mail: lnkjc@126.com
http://www.lnkj.com.cn
本书网址：www.lnkj.cn/uri.sh/7608